ATLANTIS AMBIENT AND PERVASIVE INTELLIGENCE
VOLUME 3

SERIES EDITOR: ISMAIL KHALIL

Atlantis Ambient and Pervasive Intelligence

Series Editor:

Ismail Khalil, Linz, Austria

(ISSN: 1875-7669)

Aims and scope of the series

The book series 'Atlantis Ambient and Pervasive Intelligence' publishes high quality titles in the fields of Pervasive Computing, Mixed Reality, Wearable Computing, Location-Aware Computing, Ambient Interfaces, Tangible Interfaces, Smart Environments, Intelligent Interfaces, Software Agents and other related fields. We welcome submission of book proposals from researchers worldwide who aim at sharing their results in this important research area.

All books in this series are co-published with World Scientific.

For more information on this series and our other book series, please visit our website at:

www.atlantis-press.com/publications/books

AMSTERDAM – PARIS

Multicore Systems On-Chip:
Practical Software/Hardware Design

Abderazek Ben Abdallah

University of Aizu, Adaptive Systems Laboratory,
Aizuwakamatsu, 965-8580, Fukushima-ken, Japan

ATLANTIS
PRESS

AMSTERDAM – PARIS

World Scientific

Atlantis Press

29, avenue Laumière
75019 Paris, France

For information on all Atlantis Press publications, visit our website at:

www.atlantis-press.com

Atlantis Computational Intelligence Systems

Volume 1: Agent-Based Ubiquitous Computing – Eleni Mangina, Javier Carbo, José M. Molina
Volume 2: Web-Based Information Technologies and Distributed Systems – Alban Gabillon, Quan Z. Sheng, Wathiq Mansoor

ISBN: 978-90-78677-22-2
ISSN: 1875-7669

To my wife Sonia,
my kids Tesnim and Beyram,
and my parents.

Preface

To meet high computational demands posed by latest consumer electronic devices (PDAs, cell phones, laptops, cameras, etc.), current systems employ a multitude of multicore on a single chip. The attraction of multicore processing for power reduction is compelling. By splitting a set of tasks among multiple processor cores, we can reduce the operating frequency necessary for each core, allowing to reduce the voltage on each core. Because dynamic power is proportional to the frequency and to the square of the voltage, we get a big gain, even though we may have more cores running. Even static power improves as we turn down supply voltage. However, there are several barriers to designing general purpose and embedded multicore systems. Software development becomes far more complex due to the difficulties in breaking a single processing task into multiple parts that can be processed separately and then reassembled later. This reflects the fact that certain processor jobs cannot be easily parallelized to run concurrently on multiple processing cores and that load balancing between processing cores – especially heterogeneous cores – is very difficult. The other set of problems with multicore systems design are hardware-based.

The book consists of seven chapters. The first chapter introduces multicore system architecture and describes a design methodology for these systems. The architectures used in conventional methods of MCSoCs design and custom multiprocessor architectures are not flexible enough to meet the requirements of different application domains and not scalable enough to meet different computation needs and different complexities of various applications. This chapter will emphasize on the design techniques and methodologies

Power dissipation continues to be a primary design constraint in single and multicore systems. Increasing power consumption not only results in increasing energy costs, but also results in high die temperatures that affect chip reliability, performance, and packaging cost. Energy conservation has been largely considered in the hardware design in general and also in embedded multicore system' components, such as CPUs, disks, displays, memories, and so on. Significant additional power savings can be also achieved by incorporating low-power methods into the design of network protocols used for data communication (audio, video, etc.). Chapter two investigates in details power reduction techniques at components and the network protocol levels.

Conventional on-chip communication design mostly use *ad-hoc* approaches that fail to meet the challenges posed by the next-generation Multicore Systems-on-Chip (MCSoC) designs. These major challenges include wiring delay, predictability, diverse interconnection architectures, and power dissipation.

A Network-on-Chip (NoC) paradigm is emerging as the solution for the problems of interconnecting dozens of cores into a single system-on-chip. However, there are many problems associated with the design of such systems. These problems arise from non-scalable global wire delays, failure to achieve global synchronization, and difficulties associated with non-scalable bus-based functional interconnects.

In chapter three, we explain low-power and low-cost on-chip architectures in terms of router architecture, network topology, and routing for multi- and many-core systems.

To overcome challenges from high power densities and thermal hot spots in microprocessors, multi core computing platforms have emerged as the ubiquitous computing platform from servers to embedded systems. But, providing multiple cores does not directly translate into increased performance for most applications.

With the rise of multi-core systems and many-core processors, concurrency becomes a major issue in the daily life of a programmer. Thus, compiler and software development tools will be critical to help programmers create high performance software. Chapter four covers software issues of a so-called parallelizing queue compiler targeted for future single and multicore embedded systems.

Chapter five presents practical hardware design issues of a novel dual-execution mode processor (DEP) architecture targeted for embedded applications. Practical hardware design results and advanced optimization techniques are presented in a fair mount of details

Chapter six presents design and architecture of a produced order Queue core based on Queue computing and suitable for low power computing.

With the proliferation of portable devices, new multimedia-centric applications are continuously emerging on the consumer market. These applications are pushing computer architecture to its limit considering their demanding workloads. In addition, these workloads tend to vary significantly at run time as they are driven by a number of factors such as network conditions, application content, and user interactivity. Most current hardware and software approaches are unable to deliver executable codes and architectures to meet these requirements. There is a strong need for performance-driven adaptive techniques to accommodate these highly dynamic workloads. Chapter seven shows the potential of these techniques in both software and hardware domains by reviewing early attempts in dynamic binary translation on the software side and FPGA-based reconfigurable architectures on the hardware side. It puts forward a preliminary vision for unifying runtime adaptive techniques in hardware and software to tackle the demands of these new applications. This vision will not be possible to realize unless the notorious reconfiguration bottleneck familiar in FPGAs is addressed.

Abderazek Ben Abdallah
The University of Aizu

Contents

List of Figures

List of Tables

Chapter 1

Multicore Systems Design Methodology

The strong demand for complex and high performance multicore system-on-chips (MC-SoCs) requires quick turn around design methodology and high performance cores. Thus, there is a clear need for new methodologies supporting efficient and fast design of these systems on complex platforms implementing both hardware and software modules.

This chapter starts first with an introduction about MCSoCs systems. Then a scalable core-based methodology for systematic design environment of application specific heterogeneous MCSoCs is given. The last part describes a high performance and low power synthesizable Queue core (QC-2) targeted for application specific multicore systems.

1.1 Introduction

System on chips designs have evolved from fairly simple uni-core, single memory designs to complex multicore systems on-chip consisting of a large number of IP blocks on the same silicon. As more and more cores (macros) are integrated into these designs to share the ever increasing processing load, the main challenge lies in efficiently and quickly integrating them into a single system capable of leveraging their individual flexibility. Moreover, to connect the heterogeneous cores, the multi-core architecture requires high performance complex communication architectures and efficient communication protocols architecture, such as hierarchical bus [Diefendorff (1997); Liu (2005)], point-to-point connection [Loghi (2004)], or Time Division Multiplexed Access based bus [Kulkarani (2002)].

Current design methods tend toward mixed HD/SW co-designs targeting multicore system-on-chip for specific applications [Ernst (1993); Jerraya (2005); Lennard (2000)]. To decide on the lowest cost mix of cores, designers must iteratively map the device's functionality to a particular HW/SW partition and target architecture. Every time the designers explore a different system architecture, the interfaces must be redesigned.

Unfortunately, the specific target applications generally lead to a narrow application domain and also managing all of these details is so time consuming that designers typically cannot afford to evaluate several different implementations.

Automating the interface generation is an alternative solution and a critical part of the development of embedded system synthesis tools. Currently most automation algorithms implement the system based on a standard bus protocol (input/output interface) or based on a standard component (processing) protocol. Recent work has used a more generalize model consisting of heterogeneous multicore with arbitrary communication links. The SOS algorithm [Prakash (1992)] uses an integer linear programming approach. The co-synthesis algorithm, developed in [Dave (1997)], can handle multiple objectives such as costs, performance, power and fault tolerance. Unfortunately, such design practices allow only limited automation and designers resort to manual architecture design, which is time consuming and error-prone especially in such complex SoCs.

Our design automation algorithm generates generic-architecture-template (GAT), where both processing and input/output interface may be customized to fit the specific needs of the application. Therefore, the utilization of the GAT enables a designer to make a basic architecture design without detailed knowledge of the architecture.

High performance processor cores are also needed for high performance heterogeneous multicore SoCs. Thus, we also describe a high performance synthesizable soft-core archi-tecture, which will be used as a task-distributor-core (TDC) in the MCSoC system. The system may consist, then, of multiple processing cores of various types (i.e., QueueCore(s), general purpose processor(s), domain specific DSPs, and custom hardware), and communi-cation links. The ultimate goal of our systematic design automation and architecture gener-ation is the to improve performance and the design efficiency of large scale heterogeneous multicore SoC. The rest of this work is organized as follow: Section 2 give conventional SoC design methodology. Section three gives our multicore architecture platform descrip-tion. Section four gives our core-based method for a systematic environment in a hetero-geneous MCSoC. Section five gives the synthesizable QC-2 core architecture. Section six describes the QC-2 core evaluation. In the last section we give the conclusion.

1.2   MCSoCs Design Problems

The gate densities achieved in current ASIC and FPGA devices give the designers enough logic elements to implement all the different functionalities on the same chip (SoC) by mixing self-design modules with third party one [Kulkarani (2002); Sheliga (1996); Jerraya (2005)]. This possibility opens new horizons especially for embedded systems where space constraints are as important as performance. The most fundamental characteristic of a SoC is complexity. The SoC is generally tailored to the application rather than general-purpose chip, and may contain memory, one or several specialized cores, buses, and several other digital functions. Therefore, embedded applications cannot use general-purpose computers (GPPs) either because a GPP machine is not cost effective or because it cannot provides the necessary requirements and performance. In addition, a GPP machine can't provide reliable real-time performance.

In Fig. 1.1, a typical multicore architecture is shown.
The typical model is made of a set of cores communicating through an AMBA communi-cation architecture [Diefendorff (1997)]. The communication architecture constitutes the hardware links that support the communication between cores. It also provides the system

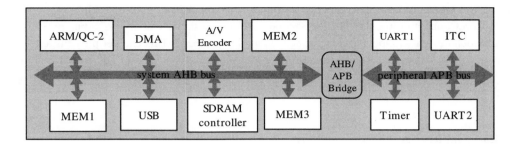

Fig. 1.1   SoC typical architecture.

with the required support for the general data transfer with external devices common to most applications. Inter-component link is often in the critical path of such a system and is a very common source of performance bottlenecks [Pasricha (2006)]. It thus becomes imperative for system designers to focus on exploring the communication design space.

Conventional SoC architectures are classified into tow types: single-processor and multi-processor architectures. Single-processor architecture consists of a single CPU and one or several ASICs. A master-slave synchronization pattern is adopted in this type. The single-processor SoC type can only offer a restricted performance capability in many applications because of the lack of true parallelism. A multiprocessor SoC architecture is a system that contains multiple instruction set processors (CPUs) and also one or several ASICs. In term of performance, multiprocessor SoCs perform better for several embedded applications. However, they (multiprocessor SoCs) introduce new challenges: first, the inter-processor communication may require more sophisticated networks than a simple shared bus; and second, the architecture may include more than one master processor. In either type, high processing performance is required because most of the applications for which SoCs are used have precise performance requirements deadlines. This is different from conventional computing, where care is generally about processing speed. We will discuss the performance issue in the following section when we introduce the QC-2 core.

In general, the architectures used in conventional methods of multiprocessor SoC design and custom multiprocessor architectures are not flexible enough to meet the requirements of different application domains (e.g. only point-to-point or shared bus communication is supported). and not scalable enough to meet different computation needs and different complexity of various applications. A promising approach was proposed in [Dave (1997)]. This method is a core-based solution, which enables integration of heterogeneous processors and communications protocols by using abstract interconnections. Behavior and communication must be separated in the system specification. Hence, system communication can be described at a higher-level and refined independently of the behavior of the system. There are two component-based design approaches: usage of a standard bus (i.e., IBM CoreConnect) protocol and usage of a standard component (i.e., VSIA) protocol [Ernst (1993); Jerraya (2005); Lennard (2000)].

For the first approach, a wrapper is designed to adapt the protocol of each component to CoreConnect protocols. For the second case, the designer can choose a bus protocol and then design wrappers to interconnect using this protocol. This chapter presents a new concepts, called virtual architecture, to cover both methods listed above. The virtual system represents an architecture as an abstract Netlist of virtual cores, which should use wrappers to get adapt accesses from the internal component to the external port.

1.3   Multicore architecture platform

The target model of our architecture consists of CPUs (i.e., QueueCore (QC-2), GPPs), hardware blocks, memories, and communication interfaces. The addition of new CPUs will not change the main principle of the proposed methodology. The processors are connected to the shared communication architecture via communication network, which maybe of whatever complexity from a single bus to a network with complex protocols. However, to ensure modularity, standard and specific interfaces to link cores to the communication architecture should be used. This gives the possibility to design separately each part of the application. For this purpose, we proposed in [Ben-Abdallah (2005)] a modular design methodology. One important feature of the proposed method is that the generic assembling scheme largely increases the architecture modularity.

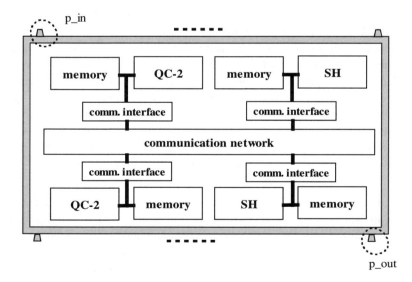

Fig. 1.2   MCSoC system platform. This is a typical instance of the architecture. In this system, the addition of a new core will not change the principle of the methodology.

Figure 1.2 show a typical instance of the platform made of 4 processors (2*QC-2 cores and 2*SH cores -Hitachi SuperH core). The QC-2 core is a special purpose synthesizable core

Task	Description
T1	Define architecture platform
T2	Describe application system level
T3	select design parameters
T4	Instantiate Pr. att.
T5	Instantiate communication
T6	mapping table
T7	Describe abstract architecture
T8	Design architecture
T9	Inst.IP cores (Pr.and Mem)
T10	H-SoC synthesis
T11	Software adaptation
T12	Binary code
T13	Pr. and Mem. emulators
T14	H-SoC validation

(described in details in section 1.5).

The designer can configure: the number of CPUs, I/O ports for each processor and inter-connections between processors, the communication protocols and the external connections (peripherals). The communication interface depends on the processor attributes and on the application-specific parameters. The communication interface that we intend to use to connect a given processor to the communication architecture, consists of two parts: one part specific to the processor's bus; the second part is generic and depends on communication protocols and on the number of communication channels used. This structure allows the "isolation" of the CPU core from the communication network.

Each interface module acts as a co-processor for the corresponding CPU. The application dependent part may include several communication channels. The arbitration is done by the CPU-dependent part and the overhead induced by this communication co-processor depends on the design of the basic components and may be very low. The use of this architecture for interfaces provides huge flexibility and allows for modularity and scalability.

1.4 Application specific MCSoC design method

In our design methodology, the application-specific parameters should be used to configure the architecture platform and an application-specific architecture is produced. These parameters result from an analysis of the application to be designed. The design flow graph (DFG) is divided into 14 "linked-tasks" as shown in Fig. 1.3 (a)-(b) and summarized in Table 1.

The first task (node T1) defines the architecture platform using all fixed architectural pa-

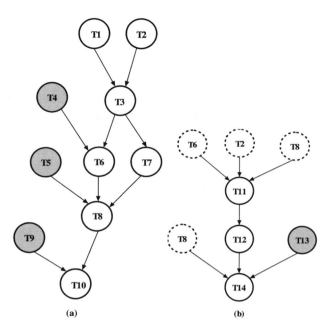

Fig. 1.3   Linked-task design flow graph (DFG). (a) Hardware related tasks, (b) Application related tasks.

rameters: (1) Network type, (2) Memory architecture, (3) CPU types, and (4) other HW modules. Using the application system level description (second task) and the architectural fixed parameters, the selection of the actual design parameters (number of CPUs, the memory sizes for each core, I/O ports for each core and interconnections, between cores, the communication protocols and the external peripherals) is performed in task 3 (node T3). The outputs of task 3 are: an abstract architecture description (node T7) and a mapping table (node T6). Node T7 is the internal structure of the target system architecture. It contains all the application specific parameters. The mapping table (T7) contains the addresses allocation and memory map for each core. The complete architecture design task (T8) is linked to the abstract architecture and the mapping table nodes (tasks). Finally, binary programs that will run on the target processors are produced in task 11 (node T11). For validation, cycle accurate simulation for CPUs and HDL (Verilog or VHDL) modeling for other cores/modules can be used for the whole architecture.

1.5   QueueCore architecture

We proposed in [Sowa (2005); ?] a produced order parallel Queue processor (QueueCore) architecture. The key ideas of the produced order queue computation model of our architecture are the operands and results manipulation schemes. The Queue computing scheme stores intermediate results into a circular queue-register (QREG). A given instruction implicitly reads its first operand from the head of the QREG, its second operand from a location explicitly addressed with an offset from the first operand location. The computed result is finally written into the QREG at a position pointed by a queue-tail pointer (QT). An important feature of this scheme is that, write after read false data dependency does not occur [Sowa (2005)]. Furthermore, since there is no explicit referencing to the QREG, it is easy to add extra storage locations to the QREG when needed. The other feature of the POC computing model is its important affect on the instruction issue hardware. The QC-1 core [?] exploits instruction-level parallelism without considerable effort and need for heavy run time data dependence analysis, resulting in a simple hardware organization when compared with conventional Superscalar processors. This also allows the inclusion of a large number of functional units into a single chip, increasing parallelism exploitation. Since the operands and result addresses of a given static-instruction (compiler generated) are implicitly *computed* during run-time, an efficient and fast hardware mechanism is needed for parallel execution of instructions. The queue processor implements a so named queue computation mechanism that calculates operands and result addresses for each instruction (discussed later). The QC-2 core, presented in this work, implements all hardware features found in QC-1 core and also supports single precision floating point accelerator.

In this section we describe the QC-2 (extended and optimized version of the QueueCore processor) architecture and prototyping results. As we explained in earlier section, the QC-2 core will be integrated into our H-SoC system.

1.5.1 *Hardware pipeline structure*

The QC-2 supports a subset of the produced order queue processor instruction set architecture [?]. All instructions are 16-bit wide, allowing simple instructions fetch and decode stages and facilitate instructions pipelining. The pipeline's regular structure allows instructions fetching, data memory references, and instruction execution to proceed in parallel. Data dependencies between instructions are automatically handled by hardware interlocks. Bellow we describe the salient characteristics of the QueueCore architecture.

(1) *Fetch (FU)*: The instruction pipeline begins with the fetch stage, which delivers four instructions to the decode unit each cycle. This is the same bandwidth as the maximum execution rate of the functional units. At the beginning of each cycle, assuming no pipeline stalls or memory wait states occur, the address pointer hardware of the fetched instructions issues a new address to the Data/Instruction memory system. This address is either the previous address plus 8 bytes or the target address of the currently executing flow-control instruction.

(2) *Decode (DU)*: The QC-2 decodes four instructions in parallel during the second phase and writes them into the decode buffer. This stage also calculates the number of consumed (CNBR) and produced (PNBR) data for each instruction [Sowa (2005)]. The CNBR and PNBR are used by the next pipeline stage to calculate source and destination locations for each instruction. Decoding stops if a queue becomes full.

(3) *Queue computation (QCU)*: The QCU calculates the first operand (*source1*) and destination addresses for each instruction. The QCU unit keeps track on the current value of the QH and QT pointers. Four instructions arrive to the QCU unit each cycle. To execute instructions in parallel, the QC-2 core must calculate the operands addresses (*source1*, *source2* and *destination*) for each instruction. Fig. 6.4 illustrates QC-2's next QH and QT pointers calculation mechanism. To calculate the *source1* address, the consumed operands (CNBR) field (port field) is added to the current QH value (QH0). The second operand address in calculated as shown in Fig. 1.5. Similar mechanism is used for the other three instructions. Because the next QH and QT values are dependent on the current QH and QT values, the calculation is performed sequentially. Each QREG entry is written exactly once and it is busy until it is written. If a subsequent instruction needs its value, that instructions must wait until it is written. After QREG entry is written, it is ready.

(4) *Barrier:* The major goal of this unit/stage is to insert barrier flags for all barrier type instructions.

(5) *Issue:* Four instructions are issued for execution each cycle. In this stage, the second operand (*source2*) of a given instruction is first calculated by adding the address *source1* to the displacement that comes with the instruction. The second operand's address calculation could be earlier calculated in the QCU stage. However, for a balanced pipeline consideration, the *source2* is calculated in this stage.

An instruction is ready to be issued if its data operands and its corresponding functional unit are available. The processor reads the operands from the QREG in the second half of stage 5 and execution begins in stage 6.

(6) *Execution (EXE)*: The macrodataflow execution core consists of 1 integer ALU unit, 1 floating-point accelerator unit, 1 branch unit, 1 multiply unit, 4 set-units, and 2 load/store units.

The load and store units share a 16-entry address window (AW), while the integer unit and the branch unit share a 16-entry integer window (IW). The FPA has its own 16-entries floating point window (FW). The load/store units have their own address generation logic. Stores are executed to memory in-order.

1.5.2   *Floating point organization*

The QC-2 floating-point accelerator (FPA) is a pipelined structure and implements a subset of the IEEE-754 single precision floating-point standard [IEEE (1985, 1981)]. The FPA consists of a floating-point ALU (FALU), floating-point multiplier (FMUL), and floating point divider (FDIV). The FALU, FMUL, FDIV and the floating-point queue-register (FQREG) employ 32-wide data paths. Most FPA operations are completed within three execution cycles. The FPA's execution pipelines are simple in design for high speeds that the QC-2 core requires. All frequently used operations are directly implemented in the hardware. The FPA unit supports the four rounding modes specified in the IEEE 754 floating point standard: round toward-to-nearest-even, round toward positive infinity, round toward

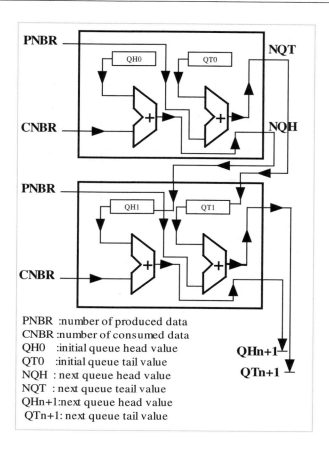

PNBR :number of produced data
CNBR :number of consumed data
QH0 :initial queue head value
QT0 :initial queue tail value
NQH : next queue head value
NQT : next queue teail value
QHn+1:next queue head value
QTn+1: next queue tail value

Fig. 1.4   Next QH and QT pointers calculation mechanism.

negative infinity, and round toward zero.

1.5.2.1   *Floating point ALU implementation*

The FALU does floating-point addition, subtraction, compare and conversion operations. Its first stage subtracts the operands exponents (for comparison), selects the larger operand, and aligns the smaller mantissa. The second stage adds or subtracts the mantissas depending on the operation and the signs of the operands. The result of this operation may overflow by a maximum of 1-bit position. Logic embedded in the mantissa adder is used to detect this case, allowing 1-bit normalization of the result on the fly. The exponent data path computes $(E + 1)$. If the 1-bit overflow occurred, $(E + 1)$ is chosen as the exponent of stage 3; otherwise, E is chosen. The third stage performs either rounding or normalization because these operations are not required at the same time. This may also result in a 1-bit overflow. Mantissa and exponent corrections, if needed, are implemented exactly in this

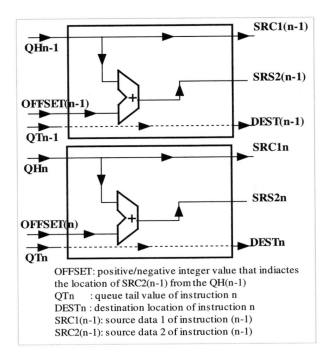

OFFSET: positive/negative integer value that indiactes
the location of SRC2(n-1) from the QH(n-1)
QTn   : queue tail value of instruction n
DESTn : destination location of instruction n
SRC1(n-1): source data 1 of instruction (n-1)
SRC2(n-1): source data 2 of instruction (n-1)

Fig. 1.5   QC-2's source 2 address calculation.

stage, using instantiations of the mantissa adder and exponent blocks.

The area efficient FADD hardware is shown in Fig. 1.6. The exponents of the two inputs (Exponent A and Exponent B) are fed into the exponent comparator, which is implemented with a subtracter and a multiplexer. In the pre-shifter, a new mantissa in created by right shifting the mantissa corresponding to the smaller exponent by the difference of the exponents so that the resulting two mantissas are aligned and can be added. The size of the preshifter is about $m * log(m)LUTs$, where m is the bit-width of the mantissa. If the mantissa adder generates a carry output, the resulting mantissa is shifted one bit to the right and the exponent is increased by one. The normalizer transforms the mantissa and exponent into normalized format. It first uses a leading-one detector (LD) circuit to locate the position of the most significant one in the mantissa. Based on the position of the LD, the resulting mantissa is left shifted by an amount subsequently deducted from the exponent. If there is an exponent overflow (during normalization), the result is saturated in the direction of overflow and the overflow flag is set. Underflows are handled by setting the result to zero and setting an underflow flag.

We have to notice that the LD anticipator can be also predicted directly from the input to the adder. This determination of the leading digit position is performed in parallel with the addition step so as to enable the normalization shift to start as soon as the addition

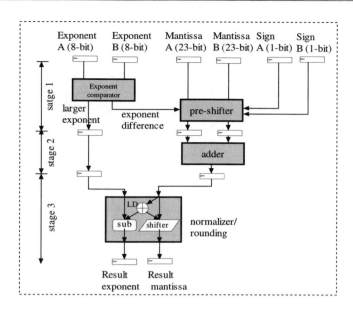

Fig. 1.6   QC-2's FADD hardware.

completes. This scheme requires more area than a standard adder, but exhibits reduced latency. For hardware simplicity and logic limitation, our FPA hardware does not support earlier LD prediction.

1.5.2.2   *Floating point multiplier implementation*

The data path of the FMUL hardware is shown in Fig. 6.7. As with other conventional architectures, QC-2's FMUL operation is much like integer multiplication. Because floating point numbers are stored in sign-magnitude form, the multiplier needs only to deal with unsigned integer numbers and normalization. Similar to the FALU, the FMUL unit is a three stages pipeline that produces a result on every clock cycle. The bottleneck of this unit was the $24 * 24$ integer multiplications.

The first stage of the floating-point multiplier is the same denormalization module used in addition to insert the implied 1 to the mantissa of the operands. In the second stage, the mantissas are multiplied and the exponents are added. The output of the module are registered. In the third stage, the result is normalized or rounded.

The multiplication hardware implements the radix-8 modified Booth [Booth (1951)] algorithm. Recoding in a higher radix was necessary to speed up the standard Booth multiplications algorithm since greater numbers of bits are inspected and eliminated during each cycle, effectively reduces the total number of cycles necessary to obtain the product. In addition, the radix-8 version was implemented instead of the radix-4 version because it reduces the multiply array in stage 2.

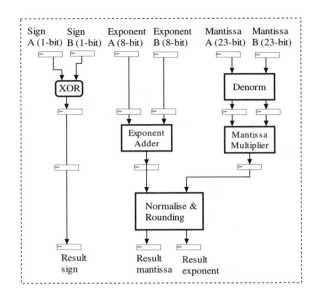

Fig. 1.7   QC-2's FMUL hardware.

1.6   QueueCore synthesis and evaluation results

1.6.1 *Methodology*

In order to estimate the impact of the description style on the target FPGAs efficiency, we have explored logic synthesis for FPGAs. The idea of this experiment was to optimize critical design parts for speed or resource optimizations.

Optimizing the HDL description to exploit the strengths of the target technology is of paramount importance to achieve an efficient implementation. This is particularly true for FPGAs targets, where a fixed amount of each resource is available and choosing the appropriate description style can have a high impact on the final resources efficiently [Micheli (2001); Gohringer (2008)]. For typical FPGAs features, choosing the right implementation style can cause a difference in resource utilization of more than an order of magnitude [Alsolaim (2000); Xilinx (2009)]. Synthesis efficiency is influenced significantly by the match of resource implied by the HDL and resources present in a particular FPGAs architecture. When an HDL description implies resources not found in a given FPGAs architecture, those elements have to be emulated using other resources at significant cost. Such emulation can be performed automatically by EDA tools in some cases, but may require changes in the HDL description in the worst case, counteracting aim of a common HDL source code base. In this work, our experiments and the results described are based on the Altera Stratix architecture [Altera (2008)]. We selected Stratix FPGAs device because it has a good tradeoffs between routability and logic capacity. In addition it has an internal embedded memory that eliminates the need for external memory module and offers up to 10

Descriptions	Modules	LE	TCF
instruction fetch unit	IF	633	414
instruction decode unit	ID	2 573	1 564
queue compute unit	QCU	1 949	1 304
barrier queue unit	BQU	9 450	4 348
issue unit	IS	15 476	7 065
execution unit	EXE	7 868	3 241
queue-registers unit	QREG	35 541	21 190
memory access	MEM	4 158	3 436
control unit	CTR	171	152
Queue processor core	QC-2	77 819	42 714

Mbits of embedded memory through the TriMatrix TM memory feature. We also used Altera Quartus II professional edition for simulation, placement and routing. Simulations were also performed with Cadence Verilog-XL tool [Cadence (2008)].

1.6.2 Design results

Figure 6.11 compares two different target implantations for 256x33 QREG for various optimizations. Depending on the target implementations device, either logic elements (LEs) or total combinational functions (TCF) are generated as storage elements. Implementations based on HardCopy device, which generates TCF functions give almost similar complexity for the three used optimizations – area (ARA), speed (SPD) and balanced (BLD). For FPGA implementation, the complexity for SPD optimization is about 17% and 18% higher than that for ARA and BLD optimizations respectively. Table 2 summarizes the synthesis

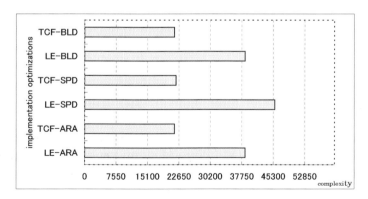

Fig. 1.8   Resource usage and timing for 256*33 bit QREG unit for different coding and optimization strategies.

results of the QC-2 for the Stratix FPGA and HardCopy targets. The complexity of each core module as well as the whole QC-2 core are given as the number of logic elements (LEs) for the Stratix FPGA device and as the TCF cell count for the HardCopy device (Structured ASIC). The design was optimized for BLD optimization guided by a properly implemented constraint table. We also found that the processor consumes about 80.4% of the total logical elements of the target device.

The achievable throughput of the 32-bit QC-2 core on different execution platforms is shown in Fig. 6.10. For the hardware platforms, we show the processor frequency. For comparison purposes, the Verilog HDL simulator performance has been converted to an artificial frequency rating by dividing the simulator throughput by a cycle count of 1 CPI. This chart shows the benefits which can be derived from direct hardware execution using a prototype when compared to processor simulation. The data used for this simulation are based on event-driven functional Verilog HDL simulation [Sowa (2005)].

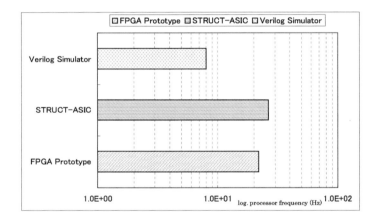

Fig. 1.9   Achievable frequency is the instruction throughput for hardware implementations of the QC-2 processor. Simulation speeds have been converted to a nominal frequency rating to facilitate comparison.

1.7   Conclusion

In this chapter we described a scalable core based methodology for generic architecture model and a Synthesizable 32-bit QC-2 core with floating point support targeted for high performance heterogeneous multicore SoC (MCSoC). The proposed design methodology is expected to have a big effect of system scalability, modularity and design time. The method also should permit a systematic generation of multicore architecture for multicore embedded system-on-chip MCSoCs. We also described the implementation and optimization of a QC-2 core processor. The 32-bit synthesizable QC-2 core supports single precision floating point support. It was correctly synthesized and tested with several Test benches. The QC-2 core was, then, optimized for speed guided by a properly implemented constraint table. We found that the processor consumes about 80.4 % of the total logical elements of the target FPGA device. It achieves about 22.5 and 25.5 MHz for 16 and 264 QREG entries respectively.

Chapter 2

Design for Low Power Systems

Power dissipation continues to be a primary design constraint in single and multicore systems. Increasing power consumption not only results in increasing energy costs, but also results in high die temperatures that affect chip reliability, performance, and packaging cost.

Energy conservation has been largely considered in the hardware design in general and also in embedded multicore systems' components, such as CPUs, disks, displays, memories, and so on. Significant additional power savings can be also achieved by incorporating low-power methods into the design of network protocols used for data communication (audio, video, etc.). This chapter investigates in details power reduction techniques at components and the network protocol levels.

2.1 Introduction

Computation and communication have been steadily moving toward embedded multicore devices. With continued miniaturization and increasing computation power, we see ever growing use of powerful microprocessors running sophisticated, intelligent control software in a vast array of devices including pagers, cellular phones, laptop computers, digital cameras, video cameras, video games, etc. Unfortunately, there is an inherent conflict in the design goals behind these devices: as mobile systems, they should be designed to maximize battery life, but as intelligent devices, they need powerful processors, which consume more energy than those in simpler devices, thus reducing battery life.

In spite of continuous advances in semiconductor and battery technologies that allow microprocessors to provide much greater computation per unit of energy and longer total battery life, the fundamental tradeoffs between performance and battery life remains critically important [Martin (1999); Tack (2004); Lafruit (2000); Lorch (1995)].

Multimedia applications and mobile computing are two trends that have a new application domain and market. Personal mobile or ubiquitous computing is playing a significant role in driving technology. An important issue for these devices will be the user interface- the interaction with its owner. The device needs to support multimedia tasks and handles many different classes of data traffic over a limited bandwidth wireless connection, including delay sensitive, real-time traffic such as video and speech.

Wireless networking greatly enhances the utility of a personal computing device. It provides mobile users with versatile communication, and permits continuous access to services and resources of the land-based network. A wireless infrastructure capable of supporting packet data and multimedia services in addition to voice will bootstrap on the success of the Internet, and in turn drive novel networked applications and services. However, the technological challenges to establishing this paradigm of personal mobile computing are non-trivial. In particular, these devices have limited battery resources. While reduction of the physical dimensions of batteries is a promising solution, such effort alone will reduce the amount of charge retained by the batteries. This will in turn reduce the amount of time a user can use the computing device. Such restrictions tend to undermine the notion of mobile computing. In addition, more extensive and continuous use of network services will only aggravate this problem since communication consumes relatively much energy. Unfortunately, the rate at which battery performance improves is very slow, despite the great interest created by the wireless business.

The energy efficiency is an issue involving all layers of the system, its physical layer, its communication protocol stack, its system architecture, its operating system, and the entire network [Lafruit (2000)]. This implicates several mechanisms that can be used to attain a high-energy efficiency. There are several motivations for energy-efficient design. Perhaps the most visible driving source is the success and growth of the portable consumer electronic market.

In its most abstract form, a networked system has two sources of energy drain required for its operation:

(1) Communication, due to energy spent by the wireless interface and due to the internal traffic between various parts of the system, and
(2) Computation, due to processing for applications, tasks required during communication, and operating system.

Thus, minimizing energy consumption is a task that will require minimizing the contributions of communication and computation.

From another had, power consumption has become a major concern because of the ever-increasing density of solid-state electronic devices, coupled with an increasing use of mobile computers and portable communication devices. The technology has thus far helped to build low power systems. The speed-power efficiency has indeed gone up since 1990 by 10 times each 2.5 years for general-purpose processors and digital signal processors (DSPs) [Tack (2004)].

Design for low-energy consumption is certainly not a new research field, and yet remains one of the most difficult as future mobile system designers attempt to pack more capabilities such as multimedia processing and high bandwidth radios into battery operated portable miniature packages. Playing times of only a few hours for personal audio, notebooks, and cordless phones are clearly not very consumer friendly. Also, the required batteries are voluminous and heavy, often leading to bulky and unappealing products [Asanovic (2006)].

The key to energy efficiency in future mobile systems will be, then, designing higher layers of the mobile system, their functionality, their system architecture, their operating system, and the entire network, with energy efficiency in mind.

2.2   Power Aware Technological-level Design optimizations

2.2.1   *Factors affecting CMOS power consumption*

Most components in a mobile system are currently fabricated using CMOS technology. Since CMOS circuits do not dissipate power if they are not switching, a major focus of low power design is to reduce the switching activity to the minimal level required to perform the computations [Najm (1994); Pedram (1996)].

The sources of energy consumption on a CMOS chip can be classified as static and dynamic power dissipation. The average power is given by:

$$P_{avg} = P_{static} + P_{dynamic} \tag{2.1}$$

The static power consumption is given by:

$$P_{static} = P_{short\text{-}ciruit} + P_{leak} = I_{sc} \cdot V_{dd} + I_{leak} \cdot V_{dd} \tag{2.2}$$

and the dynamic power consumption is given by:

$$P_{dynamic} = \alpha_{0 \rightarrow 1} \, C_L \cdot V_{dd}^2 \cdot f_{clk} \tag{2.3}$$

The three major sources of power dissipation are, then, summarized in the following equation:

$$P_{avg} = \alpha_{0 \rightarrow 1} \, C_L \cdot V_{dd}^2 \cdot f_{clk} + I_{sc} \cdot V_{dd} + I_{leak} \cdot V_{dd} \tag{2.4}$$

The first term of formula 4, represents the switching component of power, where $\alpha_{0 \rightarrow 1}$ is the node transition activity factor (the average number of times the node makes a power consuming transition in one clock period), C_L is the load capacitance and f_{clk} is the clock frequency. The second term is due to the direct-path short circuit current, I_{sc}, which arises when both the NMOS and PMOS transistors are simultaneously active, conducting current directly from supply ground. The last term, I_{leak} (leakage current), which can arise from substrate injection and subthreshold effects, is primarily determined by fabrication technology.

$\alpha_{0 \rightarrow 1}$ is defined as the average number of times in each clock cycle that a node with capacitance, C_L, will make a power consuming transition resulting in an average switching component of power for a CMOS gate to be simplified to:

$$P_{switch} = \alpha_{0 \rightarrow 1} \, C_L \cdot V_{dd}^2 \cdot f_{clk} \tag{2.5}$$

Since the energy expended for each switching event in CMOS circuits is $C_L.V_{dd}^2.f_{clk}$, it has the extremely important characteristics that it becomes quadratically more efficient as the high transition voltage level is reduced.

It is clear that operating at the lowest possible voltage is most desirable, however, this comes at the cost of increased delays and thus reduced throughput. It is also possible to reduce the power by choosing an architecture that minimizes the effective switched capacitance at a fixed voltage: through reductions in the number of operations, the interconnect capacitance, internal bit widths and using operations that require less energy per computation. We will use Formula (2.4) and (2.5) to discuss the energy reduction techniques and trade-offs that involve energy consumption of digital circuits. From these formulas, we can see that there are four ways to reduce power:

(1) reduce the capacity load C,
(2) reduce the supply voltage V,
(3) reduce the switching frequency f, and
(4) reduce the switching activity.

2.2.2 Reducing voltage and frequency

Supply voltage scaling has been the most adopted approach to power optimization, since it normally yields considerable savings thanks to the quadratic dependence of P_{switch} on V_{dd} [Najm (1994)]. The major shortcoming of this solution, however, is that lowering the supply voltage affects circuit speed. As a consequence, both design and technological solutions must be applied in order to compensate the decrease in circuit performance introduced by reduced voltage. In other words, speed optimization is applied first, followed by supply voltage scaling, which brings the design back to its original timing, but with a lower power requirement.

It is well known that reducing clock frequency f alone does not reduce energy, since to do the same work the system must run longer. As the voltage is reduced, the delay increases. A common approach to power reduction is to first increase the speed performance of the module itself, followed by supply voltage scaling, which brings the design back to its original timing, but with a lower power requirements [Pedram (1996)].

A similar problem, i.e., performance decrease, is encountered when power optimization is obtained through frequency scaling. Techniques that rely on reductions of the clock frequency to lower power consumption are thus usable under the constraint that some performance slack does exist. Although this may seldom occur for designs considered in their entirety, it happens quite often that some specific units in a larger architecture do not need peak performance for some clock/machine cycles. Selective frequency scaling (as well as voltage scaling) on such units may thus be applied, at no penalty in the overall system speed.

2.2.3 *Reducing capacitance*

Energy consumption in CMOS circuitry is proportional to capacitance C. Therefore, a path that can be followed to reduce energy consumption is to minimize the capacitance. A significant fraction of a CMOS chips energy consumption is often contributed to driving large off-chip capacitances, and not to core processing. Off-chip capacitances are in the order of five to tens of pF. For conventional packaging technologies, pins contribute approximately 13-14 pF of capacitance each (10 pF for the pad and 3-4 pF for the printed circuit board) [Borkar (1999)].

From our earlier discussion, equation (2.5) indicates that energy consumption is proportional to capacitance; I/O power can be a significant portion of the overall energy consumption of the chip. Therefore, in order to save energy, use few external outputs, and have them switch as infrequently as possible. Packaging technology can have a impact on the energy consumption. For example, in multi-chip modules where all of the chips of a system are mounted on a single substrate and placed in a single package, the capacitance is reduced. Also, accessing external memory consumes much energy. So, a way to reduce capacitance is to reduce external accesses and optimize the system by using on-chip resources like caches and registers.

2.2.3.1 *Chip layout*

There are a number of layout-level techniques that can be applied. Since the physical capacitance of the higher metal layers are smaller, there is some advantage to select upper level metals to route high-activity signals. Furthermore, traditional placement involves reducing area and delay, which in turn translates to minimizing the physical capacitance of wires. Placement that incorporates energy consumption, concentrates on minimizing the activity-capacitance product rather than capacitance alone. In general, high-activity wires should be kept short and local. Tools have been developed that use this basic strategy to achieve about 18% reduction in energy consumption.

The capacitance is an important factor for the energy consumption of a system. However, reducing the capacity is not the distinctive feature of low-power design, since in CMOS technology energy is consumed only when the capacitance is switched. It is more important to concentrate on the switching activity and the number of signals that need to be switched. Architectural design decisions have more impact than solely reducing the capacitance.

2.2.3.2 *Technology scaling*

Scaling advanced CMOS technology to the next generation improves performance, increases transistor density, and reduces power consumption. Technology scaling typically has three main goals: (1) reduce gate delay by 30%, resulting in an increase in operating frequency of about 43%; (2) double transistor density; and (3) reduce energy per transistor by about 65%, saving 50% of the power. These are not ad hoc goals; rather, they follow scaling theory [Borkar (1999)].

As the Semiconductor Industry Association roadmap (SIA) indicates, the trend of process technology improvement is expected to continue for years [SIA (1997)]. Scaling of the physical dimension involves reducing all dimensions: thus transistors widths and lengths are reduced; interconnection length is reduced, etc. Consequently, the delay, capacitance and energy consumption will decrease substantially.

Another way to reduce capacitance at the technology level is to reduce chip area. For example, an energy efficient architecture that occupies a larger area can reduce the overall energy consumption, e.g. by exploiting locality in a parallel implementation.

2.3 Power Aware Logic-level Design Optimizations

Logic-level power optimization has been extensively researched in the last few years. While most traditional power optimization techniques for logic cells focus on minimizing switching power, circuit design for leakage power reduction is also gaining importance [Ye (1998)]. As a result, logic-level design can have a high impact on the energy-efficiency and performance of the system. Issues in the logic level relate to for example state-machines, clock gating, encoding, and the use of parallel architectures.

2.3.1 *Clock gating*

Several power minimization techniques work especially well at the logic level. Most of them rely on switching frequency. The best example of which is the use of clock gating [Benini (1999)]. Clock gating provides a way to selectively stop the clock, and thus force the original circuit to make no transition, whenever the computation to be carried out by a hardware unit at the next clock cycle is useless. In other words, the clock signal is disabled to shut down some modules of the chip, that are inactive. This saves on clock power, because the local clock line is not toggling all the time.

For example the latency for the CPU of the TMS320C5x DSP processor [Benini (2001)] to return to active operation from the IDLE3 mode takes around $50\mu s$, due to the need of the on-chip PLL circuit to lock with the external clock generator. With the conventional scheme, the register is clocked all the time, whether new data is to be captured or not. If the register must hold the old state, its output is fed back into the data input through a multiplexer whose enable line (ENABLE) controls whether the register clocks in new data or recycles the existing data. However, with a gated clock, the signal that would otherwise control the select line on the multiplexer now controls the gate. The result is that the energy consumed in driving the register's clock input (CLK) is reduced in proportion to the decrease in average local clock frequency. The two circuits function identically, but utilization of the gated clock reduces the power consumption.

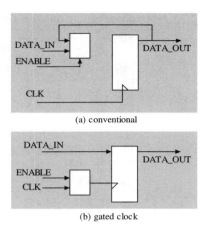

(a) conventional

(b) gated clock

Fig. 2.1   Clock gating example

2.3.2 *Logic encoding*

The power consumption can be also reduced by carefully minimizing the number of transitions. The designer of a digital circuit often has the freedom of choosing the encoding scheme. Different encoding implementations often lead to different area, power, and delay trade-offs. An appropriate choice of the representation of the signals can have a big impact on the switching activity.

The frequency of consecutive patterns in the traffic streams is the basis for the effectiveness of encoding mechanisms. For example, a program counter in a processor generally uses a binary code. On average two bits are changed for each state transition [Ben-Abdallah (2005)]. Using a Gray-code (single bit changes) can give interesting energy savings. However, a Gray-code incremental requires more transistors to implement than a ripple carry incrementer [Ben-Abdallah (2005)]. Therefore, a combination can be used in which only the most frequently changing LSB bits use a Gray code.

2.3.3 *Data guarding*

Switching activity is the major cause of energy dissipation in most CMOS digital systems. Therefore, to reduce power consumption, switching activities that do not contribute to the actual communication and computation should be eliminated. The basic idea is to identify logical conditions at some inputs to a logic circuit that is invariant to the output. Since those input values do not affect the output, the input transitions can be disabled.

Data logic-guarding technique [Tiwari (1998)], is an efficient method used to guard not useful switching activities to propagate further inside the system. The technique is based on reducing the switching activities by placing transparent latches/registers with an enable signal at the inputs of each block of the circuit that needs to be selectively turned off. If

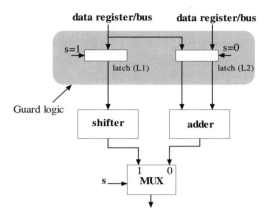

Fig. 2.2   Dual Operation ALU with Guard Logic. The multiplexer does the selection only after both units have completed their evaluation. The evaluation of one of the two units is avoided by using a guard-logic; two latches (L1 and L2) are placed with enable signals (s1 and s2) at the inputs of the shifter and the adder respectively.

the module is to be active in a clock cycle, the enable signal makes the latch transparent, permitting normal operation. If not, the latch retains its previous state and no transitions propagate through the inactive module (see Figure 2). As a summary, the logic-level design can have a high impact on the energy-efficiency and the performance of a given system. Even with the use of state of the arts hardware design language (i.e., Verilog HDL), there are still many optimizations techniques that should be explored by the designers to reduce the energy consumption at the logic-level. The most effective technique used at this level is the reduction of switching activities.

2.4   Power-Aware System Level Design Optimizations

In the previous sections we have explored sources of energy consumption and showed the low level - technology and circuit levels, design techniques used to reduce the power dissipation. In this section, we will concentrate on the energy reduction techniques at the architecture and system level.

2.4.1   *Hardware system architecture power consumption optimizations*

The implementation dependent part of the power consumption of a system is strongly related to the number of properties that a given system or algorithm may have. The component that contributes a significant amount of the total energy consumption is the communication channels or interconnects.

Experiments have already been made in designs and proved that about 10 to 40 % of the total power may be dissipated in buses, multiplexers and drivers [Liang (2004); Abnous

(1996)]. This amount can increase dramatically for systems with multiple chips due to large off-chip bus capacitance.

The energy consumption of the communication channels is largely dependent on algorithm and architecture-level design decisions. Regularity and locality are two important properties of algorithms and architectures for reducing the energy consumption due to the communication channels. The idea behind regularity is to capture the degree to which common patterns appear in an algorithm. Common patterns enable the design of less complex architecture and therefore simpler interconnect structure and less control hardware. Simple measures of regularity include the number of loops in the algorithm and the ratio of operations to nodes in the data flow graph. The statistics of the percentage of operations covered by sets of patterns is also indicative of an algorithm's regularity. Quantifying this measure involves first finding a promising set of patterns, large patterns being favored. The core idea is to grow pairs of as large as possible isomorphic regions from corresponding pairs of seed nodes [Rabaey (1995)].

Locality relates to the degree to which a system or algorithm has natural isolated clusters of operation or storage with few interconnections between them. Partitioning the system or algorithm into spatially local clusters ensures that the majority of the data transfers take place within the clusters and relatively few between clusters. The result is that the local buses with a low electrical capacity are shorter and more frequently used than the longer highly capacitive global buses. Locality of reference can be used to partition memories. Current high-level synthesis tools are targeted to area minimization or performance optimization. However, for power reduction it is better to reduce the number of accesses to long global buses and have the local buses be accessed more frequently. In a direct implementation targeted at area optimization, hardware sharing between operations might occur, destroying the locality of computation. An architecture and implementation should preserve the locality and partition and implement it such that hardware sharing is limited. The increase in the number of functional units does not necessarily translate into a corresponding increase in the overall area and energy consumption since the localization of interconnect allows a more compact layout and also fewer access to buffers and multiplexers are needed.

2.4.1.1 *Hierarchical memory system*

Efficient use of an optimized custom memory hierarchy to exploit temporal locality in the data accesses can have a very large impact on the power consumption in data dominated applications. The idea of using a custom memory hierarchy to minimize the power consumption is based on the fact that memory power consumption depends primarily on the access frequency and the size of the memory. For on-chip memories, memory power increases with the memory size. In practice, the relation is between linear and logarithmic depending on the memory library. For off chip memories, the power is much less dependent on the size because they are internally heavily partitioned. Still they consume more energy per access than the smaller on-chip memories. Hence, power savings can be obtained by accessing heavily used data from smaller memories instead of from large background memories [Su (1995,?)].

As most of the time only a small memory is read, the energy consumption is reduced. Memory considerations must also be taken into account in the design of any system. By employing an on-chip cache significant power reductions together with a performance increase can be gained. Apart from caching data and instructions at the hardware level, caching is also applied in the file system of an operating system [Su (1995)]. The larger the cache is, the better performance is achieved. Energy consumption is reduced because data is kept locally, and thus requires less data traffic. Furthermore, the energy consumption is reduced because less disk and network activity is required.

The compiler also has impact on power consumption by reducing the number of instructions with memory operands. It also can generate code that exploits the characteristics of the machine and avoids expensive stalls. The most energy can be saved by a proper utilization of registers. In [Mehta (1997)], a detailed review of some compiler techniques that are of interest in the power minimization arena is also presented.

Secondary storage

Secondary storage in modern mobile systems generally consists of a magnetic disk supplemented by a small amount of DRAM used as a disk cache; this cache may be in the CPU main memory, the disk controller, or both [Doughs (1994); Li (1994); Douglis (1994)]. Such a cache improves the overall performance of secondary storage. It also reduces its power consumption by reducing the load on the hard disk, which consumes more power than the DRAM.

Energy consumption is reduced because data is kept locally, and thus requires less data traffic. In addition, the energy consumption is reduced because less disk and network traffic is required. Unfortunately, there is trade-off in size of the cache memory since the required amount of additional DRAM can use as much as energy as a conventional spinning hard disk [Erik (1995)].

A possible technology for secondary storage is an integrated circuit called flash memory [Douglis (1994)]. Like a hard disk, such memory is non-volatile and can hold data without consuming energy. Furthermore, when reading or writing, it consumes only 0.15 to 0.47 W, far less than a hard disk. It has a read speed of about 85 ns per byte, quite like DRAM, but write speed of about $410\mu s$ per byte, about 10 to 100 times slower than hard disk. However, since flash memory has no seek time, its overall write performance is not that much worse than a magnetic disk; in fact, for sufficiently small random writes, it can actually be faster. Since flash is practically as fast as DRAM at reads, a disk cache in no longer important for read operation. The cost per megabyte of flash is about 7 to 40 times more expensive than guard disk, but about 2 to 5 times less expensive than DRAM. Thus, flash memory might also be effective as a second level cache bellow the standard DRAM disk cache [Douglis (1994); Doughs (1994)].

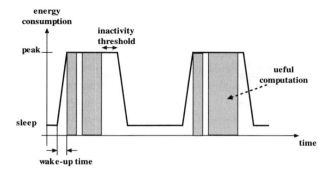

Fig. 2.3   Power consumption in typical processor

2.4.1.2   *Processor*

In general, the power consumption of the CPU is related to the clock rate, the supply voltage, and the capacitance of the devices being switched [Benini (1999); Lorch (1996); Weiser (1994); Govil (1995)]. One power-saving feature is the ability to slow down the clock. Another is the ability to selectively shut off functional units, such as the floating-point unit; this ability is generally not externally controllable. Such a unit is usually turned off by stopping the clock propagated to it. Finally, there is the ability to shut down processor operation altogether so that it consumes little or no energy. When this last ability is used, the processor typically returns to full power when the next interrupt occurs. A time energy consumption relation ships is given in Figure 2.3.

Turning off a processor has little downside; no excess energy is expended turning the processor back on, the time until it comes back on is barely noticeable, and the state of the processor is unchanged from it turning off and on, unless it has a volatile cache [Benini (1999)]. Therefore, reducing the power consumption of the processor can have a greater effect on overall power savings than it might seem from merely examining the percentage of total power attributable to the processor.

2.4.1.3   *Display and back-light*

The display and back-light have very few energy-saving features. This is unfortunate, since they consume a great deal of power in their maximum-power states; for instance, on the Duo 280c, the display consumes a maximum of 0.75 W and the back-light consumes a maximum of 3.40 [Lorch (1995,?)]. The back-light can have its power reduced by reducing the brightness level or by turning it off, since its power consumption is roughly proportional to the luminance delivered. The display power consumption can be reduced by turning the display off. It can also be reduced slightly by switching from color to monochrome or by reducing the update frequency, which reduces the range of shades or colors of Gray for each pixel, since such shading is done by electrically selecting each pixel for a particular fraction of its duty cycle. Generally, the only disadvantage of these low-power modes

Table 2.1   Operating system functionality and corresponding techniques for optimizing energy utilization

CPU scheduling	Idle power mode, voltage scaling
Operating system functionality	Energy efficient techniques
Memory allocation	Adaptive placement of memory blocks, switching of hardware energy reduction modes
Application/OS interaction	Agile content negotiation trading fidelity for power, APIs
Resource Protection and allocation	Fair distribution of battery life among both local and distributed tasks, locking battery for expensive operations
Communication	Adaptive network polling, energy-aware routing, placement of distributed computation, and server binding

is reduced readability. However, in the case of switches among update frequencies and switches between color and monochrome, the transitions can also cause annoying flashes.

2.4.2   *Operating system power consumption optimization*

Software and algorithmic considerations can also have a severe impact on energy consumption [Lorch (1998); Liang (2004); Tiwari (1994); Wolf (2002); Mehta (1997); Lorch (1995)]. Digital hardware designers have promptly reacted to the challenge posed by low-power design. Designer skills, technology improvements and CAD tools have been successful in reducing the energy consumption. Unfortunately, software engineers and system architects are often less "energy-aware" than digital designers, and they also lack suitable tools to estimate the energy consumption of their designs. As a result, energy-efficient hardware is often employed in a way that does not make optimal use of energy saving possibilities. In this section we will show several approaches to reduce energy consumption at the operating system level and to the applications.

A fundamental OS task is efficient management of host resources. With energy as the focus, the question becomes how to make the basic interactions of hardware and software as energy efficient as possible for local computation. One issue observed in traditional performance-centric resource management involves latency hiding techniques. A significant difference and challenge in energy-centric resource management is that power consumption is not easy to hide.

As one instance of power-aware resource management, we consider memory management. Memory instructions are among the more power-hungry operations on embedded processors [Tiwari (1994)], making the hardware/software of memory management a good candidate for optimization. Intels guidelines for mobile power [Intel (1998, 2001)] indicate that the target for main memory should be approximately 4 % of the power budget. This percentage can dramatically increase in systems with low power processors, displays, or without hard disks. Since many small devices have no secondary storage and rely on memory to retain data, there are power costs for memory even in otherwise idle systems. The amount of memory available in mobile devices is expanding with each new model to support more demanding applications (i.e., multimedia) while the demand for longer battery life also continues to grow significantly.

Scheduling is needed in a system when multiple functional units need to access the same object. In operating systems scheduling is applied at several parts of a system for processor time, communication, disk access, etc. Currently scheduling is performed on criteria like priority, latency, time requirements etc. Power consumption is in general only a minor criterion for scheduling; despite the fact that much energy could be saved.

Subsystems of a computer, such as the CPU, the communication device, and storage system have small usage duty cycles. That is, they are often idle and wait for the user or network interaction. Furthermore, they have huge differences in energy consumption between their operating states.

Recent advances in ad hoc networks allow mobile devices to communicate with one another, even in the absence of pre-existing base-stations or routers. All mobile devices are able to act as routers, forwarding packets among devices that may otherwise be out of communication range of one another. Important challenges include discovering and evaluating available routes among mobile devices and maintaining these routes as devices move, continuously changing the "topology" of the underlying wireless network. In applications with limited battery power, it is important to minimize energy consumption in supporting this ad-hoc communication.

There are numerous opportunities for power optimizations in such environments, including:

  i) reducing transmission power adaptively based on the distance between sender and receiver,
 ii) adaptively setting transmission power in route discovery protocols,
iii) balancing hop count and latency against power consumption in choosing the "best" route between two hosts, and
 iv) choosing routes to fairly distribute the routing duties (and the associated power consumption) among nodes in an ad-hoc network [Havinga (2001)].

2.4.3 *Application, compilation techniques and algorithm*

In traditional power-managed systems, the hardware attempts to provide automatic power management in a way that is transparent to the applications and users. This has resulted in some legendary user problems such as screens going blank during video or slide-show presentations, annoying delays while disks spin up unexpectedly, and low battery life because of inappropriate device usage. Because the applications have direct knowledge of how the user is using the system to perform some function, this knowledge must penetrate into the power management decision-making system in order to prevent the kinds of user problems described above. This suggests that operating systems ought to provide application programming interfaces so that energy-aware applications may influence the scheduling of the systems resources.

The switching activity in a circuit is also a function of the present inputs and the previous state of the circuit. Thus it is expected that the energy consumed during execution of a particular instruction will vary depending on what the previous instruction was. Thus an appropriate reordering of instructions in a program can result in lower energy. Today, the cost function in most compilers is either speed or code size, so the most straightforward way to proceed is to modify the objective function used by existing code optimizers to obtain low-power versions of a given software program. The energy cost of each instruction must be considered during code optimization. An energy aware compiler has to make a trade-off between size and speed in favor of energy reduction.

At the algorithm level functional pipelining, re-timing, algebraic transformations and loop transformations can be used [Tiwari (1994)]. The system's essential power dissipation can be estimated by a weighted sum of the number of operations in the algorithm that has to be performed. The weights used for the different operations should reflect the respective capacitance switched. The size and the complexity of an algorithm (e.g. operation counts, word length) determine the activity. Operand reduction includes common sub-expression elimination, dead code elimination etc. Strength reduction can be applied to replace energy consuming operations by a combination of simpler operations (for example by replacing multiplications into shift and add operations).

2.4.4   Energy reduction in network protocols

Up to this point we have mainly discussed the techniques that can be used to decrease the energy consumption of digital systems and focused on the computing components of a mobile host. In this subsection we will discuss some techniques that can be used to reduce the energy consumption that is needed for the communication external of the mobile host.

We classify the sources of power consumption, with regard to network operations, into two types: (1) communication related and (2) computation related.

Communication involves usage of the transceiver at the source, intermediate (in the case of ad hoc networks), and destination nodes. The transmitter is used for sending control, route request and response, as well as data packets originating at or routed through the transmitting node. The receiver is used to receive data and control packets some of which are destined for the receiving node and some of which are forwarded. Understanding the power characteristics of the mobile radio used in wireless devices is important for the efficient design of communication protocols.

The computation mainly involves usage of the CPU, main memory, the storage device and other components. Also, data compression techniques, which reduce packet length, may result in increased power consumption due to increased computation. There exists a potential trade-off between computation and communication costs. Techniques that strive to achieve lower communication costs may result in higher computation needs, and vice-versa. Hence, protocols that are developed with energy efficiency goals should attempt to strike a balance between the two costs.

Energy reduction should be considered in the whole system of the mobile and through all layers of the protocol stack. The following discussion presents some general guidelines that may be adopted for an energy efficient protocol design.

2.4.4.1 *Protocol stack energy reduction*

Data communication protocols dictate the way in which electronic devices and systems exchange information by specifying a set of rules that should a consistent, regular, and well-understood data transfer service. Mobile systems have strict constraints on the energy consumption, the communication bandwidth available, and are required to handle many classes of data transfer over a limited bandwidth wireless connection, including real time traffic such as speed and video. For example, multimedia applications are characterized by their various media streams with different quality of service requirements.

In order to save energy an obvious mode of operation of the mobile host will be a sleep mode [Sivalingam (2000)]. To support such mode the network protocols need to be modified. Store-and-forward schemes for wireless networks, such as the IEEE 802.11 proposed sleep mode, not only allow a network interface to enter a sleep mode but can also perform local retransmissions not involving the higher network protocol layers.

There are several techniques used to reduce the power consumption in all layers within the protocol stack. In Figure 2.4, we list areas in which conservation mechanisms are efficient.

Collisions should be eliminated as much as possible within the media access layer (MAC) layer, a sub layer of the data link layer, since they result in retransmissions. Retransmissions lead to unnecessary power consumption and to possibly unbounded delays. Retransmissions cannot be completely avoided in a wireless network due to the high error-rates. Similarly, it may not be possible to fully eliminate collisions in a wireless mobile network. This is partly due to user mobility and a constantly varying set of mobiles in a cell.

For example, new users registering with the base station may have to use some form of random access protocol. In this case, using a small packet size for registration and bandwidth request may reduce energy consumption. The EC-MAC protocol [Sivalingam (2000)] is one example that avoids collisions during reservation and data packet transmission. This is the default mechanism used in the IEEE 802.11 wireless protocol in which the receiver is expected to keep track of channel status through constant monitoring. One solution is to broadcast a schedule that contains data transmission starting times for each mobile as in [Sivalingam (2000)]. Another solution is to turn off the transceiver whenever the node determines that it will not be receiving data for a period of time.

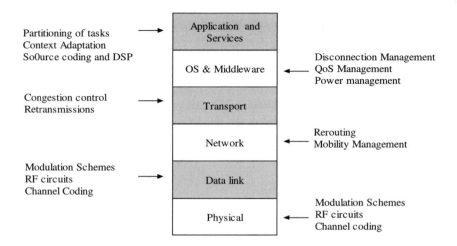

Fig. 2.4   Protocol stack of a generic wireless network, and corresponding areas of energy efficient possible research.

Physical layer

As shown in Figure 2.4, the lowest level of the protocol stack is the physical layer. This layer consists of radio frequency (RF) circuits, modulation, and channel coding systems. At this level, we need to use an energy-efficient radio that can be in various operating modes (like variable RF power and different sleep modes) such that it allows a dynamic power management [Akyildiz (2002)]. Energy can also be saved if it is able to adapt its modulation techniques and basic error-correction schemes. The energy per bit transmitted or received tends to be lower at higher bit rates. For example, the WaveLAN radio operates at 2Mb/s and consumes 1.8 W, or 0.9 J/bit. A commercially available FM transceiver (Radiometrix BIM-433) operates at 40 kb/s and consumes 60 mW, or 1.5 J/bit. This makes the low bit-rate radio less efficient in energy consumption for the same amount of data. However, when a mobile has to listen for a longer period for a broadcast or wake-up from the base station, then the high bit-rate radio consumes about 30 times more energy than the low bit rate radio. Therefore, the low bit-rate radio must be used for the basic signaling only, and as little as possible for data transfer. To minimize the energy consumption, but also to mitigate interference and increase network capacity, the transmit power on the link should be minimised, if possible.

Data link Layer

The data link layer is thus responsible for wireless link error control, security (encryption/decryption), mapping network layer packets into frames, and packet retransmission. A sub layer of the data link layer, the media access control (MAC) protocol layer is responsible for allocating the time-frequency or code space among mobiles sharing wireless channels in a region.

In an energy efficient MAC protocol the basic objective is to minimize all actions of the network interface, i.e. minimize on-time of the transmitter as well as the receiver. Another way to reduce energy consumption is by minimizing the number of transitions the wireless interface has to make. By scheduling data transfers in bulk, an inactive terminal is allowed to doze and power off the receiver as long as the network interface is reactivated at the scheduled time to transmit the data at full speed.

An example of an energy-efficient MAC protocol is E^2MaC [Havinga (1999)]. The E2MaC protocol is designed to provide QoS to various service classes with a low energy consumption of the mobile. In this protocol, the main complexity is moved from the mobile to the base station with plenty of energy. The scheduler of the base station is responsible to provide the connections on the wireless link the required QoS and tries to minimize the amount of energy spend by the mobile. The main principles of the E^2MaC protocol are avoid unsuccessful actions, minimize the number of transitions, and synchronize the mobile and the base-station.

Network layer

The network layer is responsible for routing packets, establishing the network service type, and transferring packets between the transport and link layers. In a mobile environment this layer has the added responsibility of rerouting packets and mobility management. Errors on the wireless link can be propagated in the protocol stack. In the presence of a high packet error rate and periods of intermittent connectivity of wireless links, some network protocols (such as TCP) may overreact to packet losses, mistaking them for congestion. TCP responds to all losses by invoking congestion control and avoidance algorithms. These measures result in an unnecessary reduction in the link's bandwidth utilization and increases in energy consumption because it leads to a longer transfer time.

The limitations of TCP can be overcome by a more adequate congestion control during packet errors. These schemes choose from a variety of mechanisms to improve end-to-end throughput, such as local retransmissions, split connections and forward error correction. A comparative analysis of several techniques to improve the end-to-end performance of TCP over lossy, wireless hops is given [Balakrishnan (1997)]. These schemes are classified into three categories: end-to-end protocols, where loss recovery is performed by the sender; link-layer protocols, that provide local reliability; and split-connection protocols that break the end-to-end connection into two parts at the base station. The results show that a reliable link-layer protocol with some knowledge of TCP provides good performance, more than using a split-connection approach. Selective acknowledgment schemes are useful, especially when the losses occur in bursts.

OS and middleware layer

The operating system and middleware layer handles disconnection, adaptively support, and power and QoS management within wireless devices. This is in addition to the conventional tasks such as process scheduling and file system management. To avoid the high cost, in terms of performance, energy consumption or money, of wireless network communication is to avoid use of the network when it is expensive by predicting future access and fetching necessary data when the network is cheap. In the higher level protocols of a communication system caching and scheduling can be used to control the transmission of messages. This works in particular well when the computer system has the ability to use various networking infrastructures (depending on the availability of the infrastructure at a certain locality), with varying and multiple network connectivity and with different characteristics and costs. True prescience, of course, requires knowledge of the future. Two possible techniques, LRU caching and hoarding, are for example present in the Coda cache manager. A summary of other software strategies for energy efficiency is presented in [Kistler (1993); Lorch (1998)].

2.5   Conclusion

This chapter has investigated a number of energy-aware design techniques that can be used into complex multicore systems. In particular, this chapter covered techniques used to design energy-aware systems at the technology, logic, and system levels. The vast majority of the techniques used at the system architectures, are derived from existing uni-processor energy-aware systems.

Chapter 3

Network-on-Chip for Multi- and Many-Core Systems[1]

The advance of the semiconductor technology allows us to integrate a number of IP cores such as processors, caches, and I/O modules on a single chip. To connect these cores, Network-on-Chip (NoC) that introduces a packet switched network has been widely studied and used in various types of commercial chips that include cost-effective embedded devices. Such applications often demand very tight design constraints in terms of cost and performance; thus the silicon budget available for their on-chip network should be modest. In addition, power consumption is a crucial factor, since it affects their battery life or packaging costs for heat dissipation. In this chapter, we explain low-power and low-cost on-chip architectures in terms of router architecture, network topology, and routing for multi- and many-core systems.

3.1   Introduction

The advance of the semiconductor technology allows us to integrate a number of IP cores such as processors, caches, and I/O modules on a single chip. As reported in [Pinkston (2005)], for example, as many as 1,024 functional blocks, each of which consists of a MIPS-like processor and a cache memory, will be implementable on a single chip with a 35nm technology[2]. Therefore, future on-chip interconnection networks will deal with many IP cores on a single chip, and they would be limiting factors for performance, area, and power consumption of the chip.

Interconnection networks are classified into shared-media networks or switch-media networks [Henessy (2006)]. A shared-media network transfers data on a network media (i.e., link) shared by all connected nodes, as shown in Figure 3.1. A switch-media network consists of several components: switch fabrics and point-to-point links. A switch fabric is also referred to as a router. A router has several input ports and output ports, and it dynamically establishes a connection between a set of an input port and an output port. A switch-media network can be formed by connecting routers using point-to-point links, as shown in Figure 3.2.

---

[1]Michihiro Koibuchi, National Institute of Informatics

[2]Each functional block is assumed as a single tile of Raw microarchitecture [Taylor (2002)], which has a MIPS-like processor, a 32-kByte data cache, a 96-kByte instruction cache, and on-chip routers.

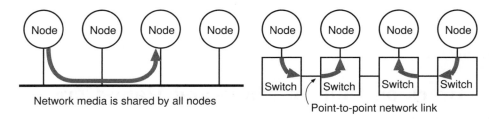

Fig. 3.1   Shared-media network   Fig. 3.2   Switch-media network

A typical example of on-chip shared-media network is an on-chip bus that connects IP cores via a shared wire link on a single chip. On-chip buses have been widely used as traditional on-chip interconnects [Flynn (1997); IBM (1999)]. Despite various techniques to improve the performance of on-chip buses [Sonics (2002); Anjo (2004); ARM (2003)], buses still create bandwidth bottlenecks when they connect a large number of IP cores, and it is difficult to increase their clock frequency because the wire delay is becoming a more severe problem in recent process technologies.

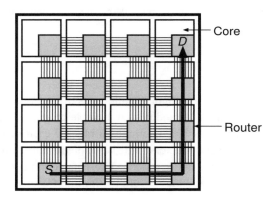

Fig. 3.3   Network-on-Chip

To avoid both bandwidth and wire-delay bottlenecks which on-chip bus structures create, on-chip switch-media networks called Networks-on-Chips (NoCs) have been studied as on-chip interconnects for the next generation [Dally (2001); Benini (2002, 2006); Vangal (2007)]. Figure 3.3 shows an example NoC that consists of sixteen tiles each of which has a processing core and a router. NoC architectures are similar to those used in massively parallel computers and System Area Networks (SANs) [Duato (2002); Dally (2004)]. In these networks, source nodes (i.e., cores) generate packets that consist of a header and a payload data. Then on-chip routers transfer them through connected links, and destination nodes decompose them. Since different packets can be simultaneously transferred on multiple links, bandwidth of NoCs is much larger than that of buses. In addition, the wire-delay problem is resolved, since each flit of a packet is transferred on limited-length point-to-point links, and buffered in every router along the routing path. By introducing error detection and re-transmission protocols, dynamic transmission errors caused by crosstalk, which will come up in future process technologies, can be also solved [Benini (2006)].

NoCs have been utilized not only for high-performance microarchitectures but also for cost-effective embedded devices mostly used in consumer equipments, such as set-top boxes or mobile wireless devices. Such embedded applications often demand very tight design con-straints in terms of cost and performance; thus the silicon budget available for their on-chip network infrastructure should be modest. On the other hand, NoCs are able to exploit the enormous wire resources, unlike inter-chip interconnects whose bandwidth is usually limited by the pin-count limitation problems outside the chip. Assuming a 0.1μm CMOS technology with 0.5μm minimum wire pitch, for example, a 3mm $\times$ 3mm tile can exploit up to 6,000 wires on each metal layer as illustrated in [Dally (2001)]. Finding the on-chip networks that effectively use large numbers of wires for low latency and high throughput communication with a modest silicon budget is thus essential for rapidly evolving embed-ded devices.

In addition, these embedded applications usually require low power, since power consump-tion is a dominant factor on their battery life, heat dissipation, and packaging cost. The overall power consumption consists of dynamic switching power and static leakage power. Switching power is still the major component of the overall power consumption during active operations; thus it should be reduced first. In addition, we need to take care of the leakage power, since it has already been consuming a substantial portion of the active power in recent process technologies, and it will further increase when the switching power is reduced after the technology is scaled down.

In this chapter, we explain low-power and low-cost on-chip architectures in terms of router architecture, network topology, and routing for multi- and many-core systems. Three ele-ments are the heart of the NoC infrastructure.

The rest of this chapter is organized as follows. Section 3.2 shows the router architecture. Section 3.3 explains the network topology, and Section 3.4 introduces the routing design for multi- and many-core systems. Section 3.5 shows the summary.

3.2 Router Architecture

The router architecture has been studied for several decades for off-chip interconnects, and various architectural innovations of routers such as wormhole switching, pipelined packet transfer, and virtual channel mechanism were developed for massively parallel comput-ers in 1980s and 90s. Although early NoCs simply have imported these architectures for on-chip purposes, various unique techniques intended to on-chip purposes are recently pro-posed [Mullins (2004, 2006); Kim (2006)].

3.2.1 *Switching Technique*

In the cases of switch-media networks, packets are transferred to their destination through multiple routers along the routing path in hop-by-hop manner. Each router forwards an incoming packet to the next router until the packet gets the final destination. The packet switching techniques decide when the router forwards the incoming packet to the neighboring router. They can be classified into three techniques: store-and-forward switching, virtual cut-through switching, and wormhole switching. They affect the network performance and buffer size needed for each router.

Store-and-Forward (SAF) Switching   The straight-forward technique for the packet switching is SAF switching, in which each router stores an entire packet in its buffer, and then it forwards the packet to the next node. That is, each router cannot forward a packet before receiving the entire packet, and it must have enough buffers that can always store the largest packet. The maximum number of cycles required to transfer a packet to its destination can be represented as $(Fh + Fb) \times D$, where D is the diameter of a given network, Fh is the number of header flits, and Fb is the number of body flits. SAF switching requires relatively large buffers in each router and its communication latency is larger than the other switching techniques described below; so SAF switching is rarely used in NoCs.

Virtual-Cut Through (VCT) Switching   In VCT switching, each router stores only fractions of a packet (i.e., flit(s)) in its buffer, and then it forwards the fractions to the next node. That is, each router can forward flits of a packet before receiving the entire packet. Thus, VCT switching has advantages in the communication latency compared to SAF switching, since the maximum number of cycles required to transfer a packet is $Fh \times D + Fb$, which is much smaller than that of SAF switching when D and/or Fb become large.

As well as SAF switching, VCT switching equips enough buffers to always store an entire packet in each router; so all fractions of a packet can be stored in a single node if the packet header cannot progress due to the conflictions. Although the communication latency of VCT switching is better than that of SAF switching, VCT switching requires large buffers for each node. As a result, VCT switching is widely used in off-chip interconnection networks like SAN.

Wormhole (WH) Switching   The most common switching technique used in NoCs is WH switching. It requires small buffers for each router so as to at least store a header flit in each hop, and each router can forward flits of a packet before receiving the entire packet. Thus, WH switching has advantages in the communication latency compared to SAF switching as well as VCT switching, though its buffer requirement is much smaller than that of VCT switching. This is the reason why WH switching is widely used in NoCs.

In WH switching, flits in a packet are often stored across multiple routers along the routing path, and their movement much looks like a "worm".

Again, fractions of a packet can be stored across different nodes along the routing path; so the single packet often keeps occupying buffers in multiple routers along the path when the header of the packet cannot progress due to the conflictions. Such situation is referred to as "Head-of-Line (HoL) blocking". The occupied buffers due to HoL blockings prevents other packet transfers that go through these lines. The problem of WH switching is the performance degradation due to the frequent HoL blockings.

WH Switching with Virtual Channels To equip multiple buffers in a single physical channel is the most common way to mitigate the HoL blockings of WH switching. In this case, each of multiplexed buffers in a physical channel acts as a "virtual" channel [Dally (1992)], and incoming packets would progress if more than one virtual channels in a physical channel are not occupied. Therefore, the occurrence of HoL blockings will be reduced as the number of virtual channels multiplexed in a physical channel increases.

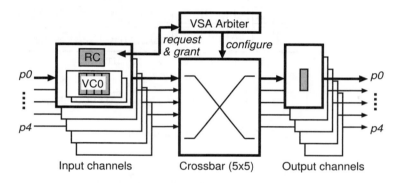

Fig. 3.4   Wormhole router architecture

3.2.2   *Router Components*

Figure 3.4 illustrates a conventional wormhole router with virtual channels. This router consists of a crossbar switch, an arbitration unit, and five input physical channels (or ports), and it is likely to keep components to a minimum for packet switching. We use it as a baseline router architecture for ease of understanding, and we explain each component in this section.

Physical Channel Each physical channel consists of a routing computation (RC) unit and virtual channels, each of which has a FIFO buffer for storing for flits.

Input Buffer This router has buffers at only its input channels. These FIFO buffers can be implemented with either SRAMs or registers, depending on the depth of the buffers, not the width.

Crossbar Switch To mitigate the HoL blocking problem, a virtual-channel design sometimes employs a $pv \times pv$ full crossbar, where p is the number of physical channels and v is the number of virtual channels. However, the crossbar complexity significantly increases when using the $pv \times pv$ crossbar. In addition, its performance improvement will be limited because the data rate out of each input port is limited by its bandwidth [Dally (1992)]. Therefore, an on-chip router usually employs a small $p \times p$ crossbar by just duplicating the buffers.

3.2.3 Pipeline Processing

We explain a pipeline structure of the baseline router shown in Figure 3.4.

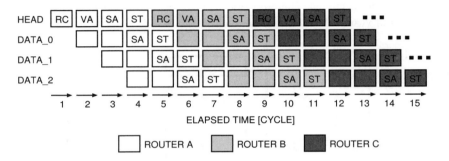

Fig. 3.5   Router pipeline structure (4-cycle)

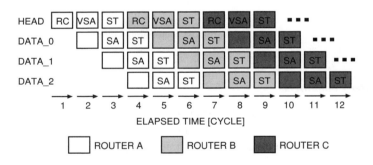

Fig. 3.6   Router pipeline structure (3-cycle)

Packet is transferred using a pipeline processing which can be simply split into four stages in the router. The 4-cycle processing quoted from [Dally (2004)] is used as a baseline as shown in Figure 3.5.

In the router, a header flit is transferred through four pipeline stages that consist of the routing computation (RC) stage, virtual channel allocation (VA) stage for output channels, switch allocation (SA) stage for allocating the time-slot of the crossbar switch to the output channel, and switch traversal (ST) stage for transferring flits through the crossbar.

Packet-switched network, which relies on routers, is more complex than that of on-chip

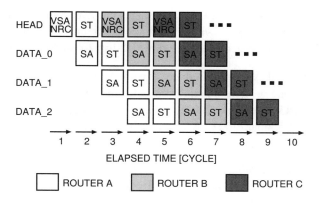

Fig. 3.7   Router pipeline structure (2-cycle)

bus. The complex router introduces the difficulty in providing low latency, low power, and high reliability. Thus, advanced on-chip routers have been widely researched for improving them.

3.2.3.1   *Toward Low Latency*

Here, we introduce some low-latency techniques of the pipelined routers.

Speculative Router   A well-known technique to reduce the number of pipeline stages in a router is the speculative transfer that performs different pipeline stages in parallel [Peh (2001)]. Figure 3.6 shows an example of the speculative router that performs the VA and SA in parallel. These operations are merged into a single stage, called the virtual channel and switch allocation (VSA) stage. Notice that when the VA operation in the VSA stage cannot be completed due to the conflicts with other packets, the SA operation also fails regardless of its result and must be re-executed. For further reducing the pipeline stages, the double speculation that performs RC, VA, and SA operations in parallel is possible, but it would degrade the performance due to the frequent miss speculations and retries.

Look-Ahead Router   The look-ahead routing technique removes the control dependency between the routing computation (RC) and switch allocation (SA) in order to perform them in parallel, by selecting the output channel of the i-th hop router in the $(i-1)$-th hop router [Galles (1996)]. That is, each router performs the routing computation for the next hop (denoted as NRC), as shown in Figure 3.7. Since the computational result at the NRC stage of the $(i-1)$-th hop router is used in the i-th hop router, the result does not affect the following VA and SA operations in the $(i-1)$-th hop router; therefore the NRC and VSA operations can be performed in parallel (Figure 3.7).

However, the NRC stage should be completed before the ST stage in the same router, because the hint bits in a packet header, which are the results of the NRC, must be updated for the NRC/VSA stage of the next router before the packet header is sent out. Thus, the control dependency between the NRC and ST stages in a router still remains difficulty for shortening to a single cycle router without harming the frequency, although some aggressive attempts using this approach have been done [Dally (2004); Mullins (2004, 2006)].

Bypassing Router   This section introduces existing aggressive low-latency router architectures that bypass one or more pipeline stages for the specific packet transfers, such as paths frequently used so far [Park (2007)], packets continually moving along the same dimension [Izu (1994); Kumar (2007); Krishna (2008)], and paths pre-specified in response to the application [Michelogiannakis (2007); Koibuchi (2008)].

Mad-postman switching exploits the path regularity of dimension-order routing on meshes for reducing the communication latency on off-chip networks that use bit-serial physical channels [Izu (1994)]. In Mad-postman switching, a router forwards an incoming packet to the output channel in the same dimension as soon as it receives the packet.

Express virtual channels (EVCs) have been proposed to enable some network packets to entirely bypass the RC, VA, and SA within a single dimension of the on-chip routers [Kumar (2007); Krishna (2008)]. Virtual bypassing paths are established between non-adjacent routers on the same dimension, and the flits can be transferred at the minimal latency (e.g., 1-cycle per a hop) in the intermediate routers. This method is efficient for reducing the communication latency of long-haul packet transfers that span multiple intermediate routers.

Dynamic fast path architecture reduces the communication latency of frequently-used paths by sending a switch arbitration request to the next node before flits in the fast path actually enter the next node [Park (2007)].

Preferred path employs a bypass datapath that connects input and output channels without using the crossbar switch in a router, in addition to the original datapath that goes through the crossbar [Michelogiannakis (2007)]. The bypass datapath can be customized so as to reduce the communication latency between the specific source-destination pairs. Also, Default-backup path (DBP) mechanism provides a low-latency unicursal ring network that spans all routers on a chip, though it has been originally proposed for on-chip fault-tolerance [Koibuchi (2008)]. All techniques listed above can bypass some pipeline stages only for the specific packet transfers, based on a static or a single bypassing policy.

Prediction Router   As yet another low-latency router architecture, the authors have proposed the prediction router that predicts an output channel being used by the next packet transfer and speculatively completes the switch arbitration [Matsutani (2009)]. In the prediction routers, incoming packets are transferred without waiting the routing computation and switch arbitration if the prediction hits. Otherwise, they are transferred through the original pipeline stages without any additional latency overheads.

Figure 3.8 illustrates the prediction router architecture changed from the original one mentioned above. The changes required for the prediction router are as follows:

(1) adding a predictor for each input channel,

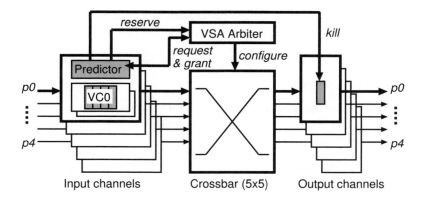

Fig. 3.8   Prediction router architecture

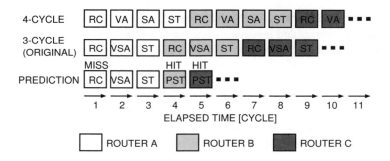

Fig. 3.9   Prediction router pipeline

(2) changing the arbitration unit so that it can handle the tentative reservations from predictors, and

(3) adding the *kill* signal for each output channel in order to remove mis-routed flits when the prediction fails.

Figure 3.9 shows a timing diagram of a packet transfer from router (a) to router (c) via router (b) in the cases of the 3-cycle speculative router (denoted as original) and the prediction router. In the prediction router, the prediction hits in router (b) and router (c), while it fails in router (a).

Since the prediction router completes the RC, VA, and SA operations prior to packet arrivals, it performs the predictive switch traversal (PST) as soon as a new packet arrives; thus the header flit transfer is completed in a single cycle without waiting the RC and VSA stages if the prediction hits. Otherwise, although dead flit(s) are forwarded to the wrong output channel, their propagation is stopped inside a router and their copy is forwarded to the correct output channel through the original pipeline stages (i.e., RC, VSA, and ST stages) without any additional latency overheads. Notice that the predictive switching can be applied to various router architectures, such as the look-ahead 2-cycle router.

The primary concern for reducing the communication latency is the hit rates of prediction algorithms, and up to 90% of the prediction hit rate is achieved in real applications [Matsutani (2009)].

3.2.3.2 Toward Low Power

Various low-power techniques have been proposed for microprocessors, and they have also been applied to on-chip routers. In this section, we show low-power techniques for microprocessors, off-chip routers, and on-chip routers.

Reducing Switching Activity   The clock gating technique shuts off the clock signals to registers and latches when output signals from these storage elements are ignored [Moyer (2001)]. The operand isolation technique selectively shuts off the propagation of the switching activity incurred by redundant operations [Munch (2000)]. Both techniques reduce the redundant switching activity of circuit blocks; thus they can save the dynamic switching power of the blocks. In [Mullins (2006)], Mullins analyzed the effect of clock gating on on-chip routers and he demonstrated the significant benefits of clock gating for them. These techniques are very common in on-chip router implementations, and they have already been applied to various router designs.

Voltage and Frequency Scaling   The voltage and frequency scaling is a power saving technique that reduces the operating frequency and the supply voltage according to the applied load. The dynamic power consumption is related to the square of the supply voltage; thus, because a peak performance is not always required during the whole execution time in most cases, adjusting the frequency and supply voltage so as to at least achieve the required performance can reduce the dynamic power.

The voltage and frequency scaling techniques have been applied to microprocessors [Nakai (2005); Nowka (2002)], accelerators [Kawakami (2006)], and network links [Shang (2003); Stine (2004)]. In [Shang (2003)], the frequency and the voltage of network links are dynamically adjusted based on the past utilization. In [Stine (2004)], the network link voltage is scaled down by distributing the traffic load using an adaptive routing.

The voltage and frequency scaling techniques can be classified into two schemes: *dynamic* and *static*. The frequency can be controlled by a PLL frequency divider, and the supply voltage can be adjusted by controlling an off-chip dc-dc converter [Nakai (2005)]. They are adaptively adjusted in the dynamic scheme, while they are statically configured at the beginning of each application in the static scheme. The transition time of the clock rate and the supply voltage cannot be negligible in the dynamic scheme (e.g., 10,000 cycles [Stine (2004)] or $50 \mu s$ [Kawakami (2006)]); so the frequent transitions sometimes overwhelm the benefits of dynamic scheme as reported in [Stine (2004)].

Power Gating   Power gating is a representative leakage-power reduction technique, which shuts off the power supply of idle blocks by turning off (or on) the power switches inserted between the VDD line and the blocks or between the GND line and the blocks.

This concept has been applied to circuit blocks with various granularities, such as processor cores [Ishikawa (2005)], execution units in a processor [Hu (2004)], and primitive gates [Usami (2006)].

We need to understand both the negative and positive impacts of power gating when we use it. Actually, a state transition between the sleep and active modes incurs a performance penalty, and turning the power switches on or off consumes an overhead energy, which means a short-term sleep rather increases the power consumption. In [Hu (2004)], an analytical model of the run-time power gating of the execution units in a microprocessor is provided. The following three parameters quoted from [Hu (2004)] affect the performance and energy.

- T_{wakeup}: Number of cycles required to charge up the local voltage of a sleeping block. A delay for turning on its power switch is also lumped into the T_{wakeup} value.
- $T_{\text{idledetect}}$: Number of cycles required to detect an idle duration in an active block and decide to shut off the block. A delay for turning off its power switch is also lumped into the $T_{\text{idledetect}}$ value.
- $T_{\text{breakeven}}$: Number of sleep cycles at least required to compensate for the overhead energy to turn the power switch on and off.

The T_{wakeup} value affects the performance (e.g., packet throughput of routers), since a pipeline stall will occur if a new request suddenly comes to a sleeping block. Also, $T_{\text{idledetect}}$ shortens the sleep duration of blocks, since an idle block must stay in the active state for $T_{\text{idledetect}}$ cycles before it decides to go to the sleep mode.

A short-term sleep of less than $T_{\text{breakeven}}$ cycles cannot compensate for the energy overhead of driving a power switch, and the power consumption will be increased; thus the $T_{\text{breakeven}}$ value determines the benefits of power gating. The $T_{\text{breakeven}}$ value depends on various parameters, such as the sizes of a power switch and a decoupling capacitance. Hu *et al.* report that $T_{\text{breakeven}} \approx 10$ based on the typical parameters of a recent microprocessor [Hu (2004)].

Dynamic Link Shutdown   Chen *et al.* propose power-aware router buffers based on Drowsy and Gated V_{dd} SRAMs to regulate their leakage power [Chen (2003)]. Soteriou *et al.* provide a thorough discussion about power-aware networks whose links can be turned on and off, in terms of connectivity, routing, wake-up and sleep decisions, and router pipeline architecture [Soteriou (2004, 2007)].

These works are proposed for both off-chip and on-chip interconnects, and they assume to use relatively large buffers in their routers. In [Chen (2003)], the router buffers are constructed with SRAMs. As a sleep control policy for the buffer, a certain portion (i.e., window size) of the buffer is made active before it is accessed. By tuning the window size, the input ports can always provide enough buffer space for the arrival of packets, and the network performance will never be affected [Chen (2003)]. On the other hand, in the case of low-cost lightweight routers, since the buffer depth is shorter than the window size, the wake-up delay of the buffers directly affects the network performance if the links or channels are dynamically turned on and off. A state transition between sleep and active mode incurs the performance penalty, and turning a power switch on or off dissipates the overhead energy, which means a short-term sleep adversely increases the power consumption. Thus, a sleep control method based on look-ahead routing that detects the arrival of packets two hops ahead has been proposed in order to hide the wake-up delay and reduce the short-term sleeps of channels [Matsutani (2008)].

3.2.3.3 *Toward High Reliability*

The dependability is another crucial factor in design of on-chip routers. Here, we focus on a technique to tolerate transient soft failure that causes data to be momentarily discarded (e.g., bit error), and this loss can be recovered by a software layer.

Error control mechanism can work for each flit or each packet. If the error rate is high, an end-to-end flow control makes the large number of retransmitted packets that increase both power consumption and latency. To mitigate the number of retransmitted packets, an error correction mechanism is employed to every packet or flit, in addition to error detection. Instead of an end-to-end flow control, a switch-to-switch flow control can be used, and it checks whether a flit or packet includes error, or not at every switch. Thus, a switch-to-switch flow control achieves the smaller average latency of packets compared with that of end-to-end flow control, although each switch additionally has retransmission buffers that would increase the total power consumption. A high-performance error control mechanism, such as error correction at switch-to-switch flow control, can provide high-reliable communication with small increased latency even at the high error-rate. A sustainable low voltage usually increases the bit error rate, but it could be recovered by a high-performance error control mechanism. Thus, they could contribute to make a low-power chip, although the power/latency trade-off of the error control mechanism is sensitive. The best error recovery schemes depends on various design factors, such as flit-error rate, link delay, and packet average hops in terms of latency and power consumption [Murali (2005)].

In addition to transient soft failure that causes data to be momentarily, hard permanent failure could occur. A link has wide bit-widths. It is possible to employ channel reconfiguration techniques at router to tolerate link faults with graceful performance degradation by decoupling singular wire faults from the other wires composing the channel [Dally (2004)].

Router architectures have also been proposed that include fault-tolerant mechanisms for bypassing hard faults in some units along the router internal datapath such as the routing computation, input buffer, and switch arbiter units [Kim (2006)]. Similarly, a mechanism called BLAM provides central bypass buffers so that a packet passes the previous misrouted packet on the same input port [Thottethodi (2003)]. Another technique using additional datapaths is called "preferred paths" in order to drastically reduce packet latency at routers [Michelogiannakis (2007)]. These techniques, while useful for the purposes proposed, do not provide support for bypassing the entire router internal datapath nor network-wide support for ensuring connectivity and deadlock-free routing on those resources, while the Default-backup path (DBP) mechanism can perform it [Koibuchi (2008)]. The DBP mechanism uses nominal redundancy to maintain network connectivity of non-faulty NoC routers and healthy on-chip PEs in the presence of hard failures occurring in the network. In achieving a lightweight reliable structure, the mechanism provides nominal default paths as backup between certain router ports which serve as alternative datapaths to circumvent failed components (i.e., input buffers, crossbar switch, etc.) within a faulty NoC router, as shown in Figure 3.10.(a) (the bypassing path from input port $y-$ to output port $y+$ at the black router in Figure 3.10.(b)).

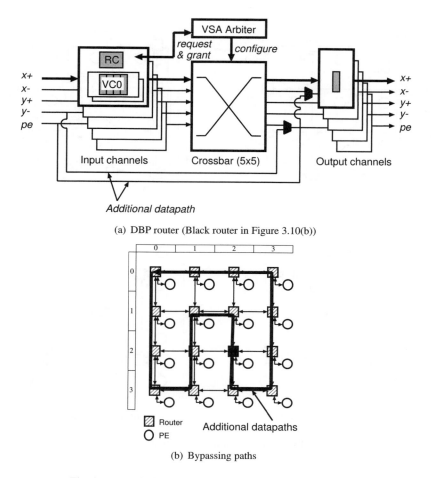

(a) DBP router (Black router in Figure 3.10(b))

(b) Bypassing paths

Fig. 3.10   Default-backup path mechanism in 16-core NoC

Along with a minimal subset of normal network channels, the set of default backup paths internal to faulty routers form—in the worst case—a unidirectional ring topology that provides network-wide connectivity to all PEs. This lightweight fault-tolerant mechanism is premised on the notion that complicated redundancy techniques need not be used to support high reliability in NoCs.

3.3 Topology

In this section, we introduce various network topologies which define the connection pattern between routers and cores in a network. Then we analyze their characteristics.

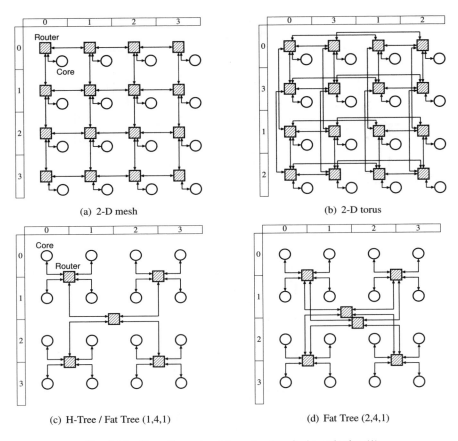

(a) 2-D mesh

(b) 2-D torus

(c) H-Tree / Fat Tree (1,4,1)

(d) Fat Tree (2,4,1)

Fig. 3.11   Two-dimensional layouts of typical topologies (1)

3.3.1 Interconnection Topologies

Figures 3.11 and 3.12 show typical on-chip networks, where a white circle represents a processing core and a shaded square represents a router connecting other routers or cores.

Figure 3.11(a) shows a two-dimensional mesh topology. Mesh topologies can be repre-

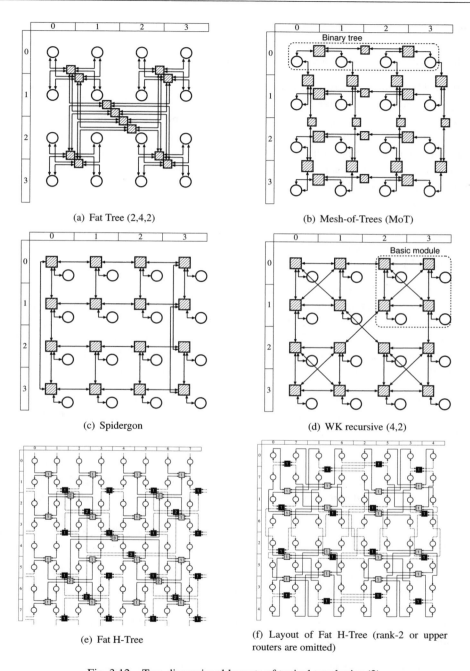

(a) Fat Tree (2,4,2)

(b) Mesh-of-Trees (MoT)

(c) Spidergon

(d) WK recursive (4,2)

(e) Fat H-Tree

(f) Layout of Fat H-Tree (rank-2 or upper routers are omitted)

Fig. 3.12   Two-dimensional layouts of typical topologies (2)

sented as k-ary n-mesh, where k is the number of nodes in a dimension (i.e., array size) and n is the number of dimensions. Although its node degree is five, those of edge nodes and corner nodes are four and three, resulting in a nonsymmetric form.

Figure 3.11(b) shows a two-dimensional torus topology, which has wrap-around channels that connect nodes in top edge and bottom edge or nodes in left edge and right edge. Thus, a torus is a mesh with wrap-around channels. Torus topologies can be represented as k-ary n-cube, where k is the array size and n is the number of dimensions. Since all nodes in a torus have five ports, torus is a symmetric topology, unlike mesh that does not have wrap-around ports. A torus has twice bisection bandwidth of a same-sized mesh due to its wrap-around channels. In addition, the average hop count of a torus is shorter than that of the same-sized mesh.

Two-dimensional mesh and torus have been employed as a typical on-chip interconnects, because their grid-based regular arrangements are intuitively considered to match the two-dimensional VLSI layout.

On the other hand, constant attention has been focused on tree-based topologies, because of their relatively short hop-count, which enables a lower latency communication than mesh and torus. The simplest tree-based topology is H-Tree in which each router (except for the top-rank router) has one upward and four downward connections (Figure 3.11(c)). Thus, the number of ports is five in every router except the root. However, the links and routers around the root of the tree are frequently congested due to its poor bisection bandwidth.

To mitigate the congestion around the root of the tree, Fat Tree enhances the number of connections toward the root [Leiserson (1985)]. As stylized in [Leiserson (1985); DeHon (2004)], various forms of Fat Tree can be created, and they can be expressed with a triple (p, q, c), where p is the number of upward connections, q is the number of downward connections, and c is the number of upward connections that each core has. Figure 3.12(a) shows a typical Fat Tree $(2,4,2)$, in which each router (except for top-rank routers) has two upward and four downward connections, and each core has two upward connections. This is the network architecture used in CM-5 [Leiserson (1996)]. Figure 3.11(d) shows a smaller version labeled as $(2,4,1)$, which means every core has only one upward connection. Note that a Fat Tree $(1,4,1)$ is identical to the H-Tree.

With a large p and q, the total bandwidth of Fat Tree is enlarged. However, the degree of a router becomes $p+q$ and so the hardware requirements for the router and its wires are also increased. Using a large c can improve the bisection bandwidth of Fat Tree almost linearly. That is, the bisection bandwidth can be doubled when c is doubled. However, the number of routers is almost doubled and it requires a lot of routers and wires between them.

A Mesh-of-Trees (MoT) network [Leighton (1984); DeHon (2004)] has properties of both mesh and tree. In an MoT network, starting with a mesh of processing cores, the cores in each row and column of the mesh are connected by a tree, as shown in Figure 3.12(b). It can utilize the traffic locality.

The butterfly network (*k*-ary *n*-fly) [Dally (2004)] can be efficiently mapped onto a 2-D VLSI by utilizing high-radix routers [Kim (2005)]. In the flattened butterfly [Kim (2007)], routers in each row of a conventional butterfly are combined into a single router. Notable improvements, such as adding bypass channels for non-minimal routing with minimal increase in power, have been proposed for it [Kim (2007)].

The Spidergon topology, which is a bidirectional ring with additional channels that connect diagonal counterparts in the ring, has been proposed for cost-effective on-chip networks [Coppola (2004)]. It can be efficiently mapped onto a chip, as shown in Figure 3.12(c). Although it has a good cost-performance property, its average hop count considerably increases as the number of nodes increases, even though it provides diagonal links to mitigate the increase of diameter compared to a conventional ring.

The 2-D shuffle-exchange mesh (SEM) [Nadooshan (2008)] is formed by applying the conventional shuffle-exchange structure in each row and each column of a network. Compared with a traditional mesh, it can reduce the diameter with the same cost by replacing some links with nonadjacent connections based on the shuffle and exchange operations. The optimal 3-D layout scheme of 2-D SEM that can reduce nearly 30 % of the link power has also been proposed [Sharifi (2008)].

In addition, the WK(*d*, *t*) network [Vecchia (1988)] (Figure 3.12(d)) has been researched for on-chip purposes [Rahmati (2006)]. It can be constructed by grouping basic modules, each of which is a *d*-node complete graph, recursively for *t* times. Thorough comparisons between WK recursive and mesh are provided in [Rahmati (2006)]. Although a WK recursive provides lower latency than a same-sized mesh for low traffic loads, its saturated throughput is comparable or inferior to the mesh.

A Fat H-Tree [Matsutani (2007)] has a torus structure, which is formed by combining only two H-Trees (B-tree and R-tree in Figures 3.12(e) and 3.12(f)), and it offers performance comparable to the torus by using a smaller network logic. Similar to the Fat Tree (2,4,2) in Figure 3.12(a), every processing core in a Fat H-Tree has two ports: one for connecting to the red tree and the other one for the black tree. Interconnection networks that have both tree and grid structures have been researched for large-scale parallel machines. The Recursive Diagonal Torus (RDT) [Yang (2001)] is an extended hierarchical torus that also has the tree property. However, since RDT was originally designed for massively parallel machines, its node degree is high (e.g., at least 8), so its connection structure tends to be costly in a microarchitecture domain and its layout on a chip is also difficult.

A large number of network topologies have been proposed so far, but those employed in practical systems are limited to some well-known topologies, such as 2-D mesh, torus, and Fat Trees; therefore we mainly compare with 2-D mesh, torus, and Fat Trees in this section.

Table 3.1   Channel bisection B_c

	N-core	16-core	64-core	256-core
H-Tree	4	4	4	4
Fat Tree (2,4,1)	2^{n+1}	8	16	32
Fat Tree (2,4,2)	2^{n+2}	16	32	64
Fat H-Tree	$2^{n+2}+8$	24	40	72
2-D Mesh	2^{n+1}	8	16	32
2-D Torus	2^{n+2}	16	32	64

Table 3.2   Average hop count H_{ave}

	16-core	64-core	256-core
H-Tree, Fat Tree(2,4,*)	3.60	5.43	7.36
Fat H-Tree	3.20	4.84	6.88
2-D Mesh	4.67	7.33	12.67
2-D Torus	4.13	6.06	10.03

3.3.2 Topological Properties

3.3.2.1 Ideal Throughput

The ideal throughput of a network is the data acceptance rate that would result from a perfectly balanced routing and a flow control with no idle cycles; it is calculated as [Dally (2004)]

$$\Theta_{ideal} \leq \frac{2bB_c}{N}, \tag{3.1}$$

where N is the number of cores, b is the channel bandwidth, and B_c is the channel bisection of the network. Table 3.1 shows the channel bisection of typical networks. Again, the number of cores is $N = 2^n \times 2^n$; therefore the number of ranks in an N-core tree is $\log_4(N) = \log_4(4^n) = n$.

3.3.2.2 Average Hop Count

The number of source-destination pairs in an N-core network is $N^2 - N$; thus the average hop count in the network is

$$H_{ave} = \frac{1}{N^2 - N} \sum_{x,y \in N} H_{(x,y)}, \tag{3.2}$$

where $H_{(x,y)}$ is the hop count from core-x to core-y. Table 3.2 shows the average hop count of typical networks for uniform random traffic, in which each source sends equally to each destination. The average hop count depends on whether the routing includes non-minimal paths.

Note that the average hop count varies depending on the traffic pattern and task mapping.

Table 3.3   Number of routers R

	N-core	16-core	64-core	256-core
H-Tree	$(4^n - 1)/3$	5	21	85
Fat Tree (2,4,1)	$(4^n - 2^n)/2$	6	28	120
Fat Tree (2,4,2)	$4^n - 2^n$	12	56	240
Fat H-Tree	$2(4^n - 1)/3$	10	42	170
2-D Mesh	N	16	64	256
2-D Torus	N	16	64	256

3.3.2.3   Number of Routers

The number of routers in a chip affects the network logic area and the implementation cost. In the case of mesh and torus topologies, the number of routers is obvious as shown in Table 3.3.

In the case of H-tree, it can be calculated as follows. Assuming that the number of downward connections q is four, the number of routers in an n-rank H-Tree network, R_{ht}, is

$$R_{ht} = (q^n - 1) / (q - 1) = (4^n - 1) / 3. \tag{3.3}$$

Similarly, the number of routers in a Fat Tree (2,4,1) network, R_{ft1}, is

$$R_{ft1} = (q^n - 2^n) / (q - 2) = (4^n - 2^n) / 2. \tag{3.4}$$

A Fat Tree (2,4,2) contains twice as many routers as are in the Fat Tree (2,4,1); therefore the number of routers in the Fat Tree (2,4,2) is $R_{ft2} = 2R_{ft1}$.

Table 3.3 lists the number of routers in typical networks. The number of routers in a Fat H-Tree is less than that in a Fat Tree (2,4,2), yet the Fat H-Tree outperforms the Fat Tree in terms of ideal throughput and average hop count, as shown in Section 3.3.2.1 and 3.3.2.2.

3.3.2.4   Total Unit-Length of Links

We present the wire length of typical topologies in the case of 2-D layout. Assuming that the distance between two neighboring cores aligned in a 2-D grid square is 1-unit, we define L as the total unit-length of links in a given network. For instance, the 16-core H-Tree network shown in Figure 3.11(c) has 16 1-unit-length links and four 2-unit-length links, thus its L is 24-unit.

In the case of mesh and torus topologies, the total unit-length of links is obvious.

The total unit-length of links in an n-rank H-Tree network, $L_{2D,ht}$, can be expressed as

$$L_{2D,ht} = \sum_{i=1}^{n} l_{ht}^i \cdot r_{ht}^i, \tag{3.5}$$

where l_{ht}^i is the total unit-length of links between a rank-i router and its four child routers, and r_{ht}^i is the number of rank-i routers in the H-Tree. Assuming that the number of cores

Table 3.4 Total unit-length of links L (1-unit=distance between neighboring two cores)

	N-core	16-core	64-core	256-core
H-Tree	$2(N - 2^n)$	24	112	480
Fat Tree (2,4,1)	nN	32	192	1,024
Fat Tree (2,4,2)	$2nN$	64	384	2,048
Fat H-Tree	$8 + 8(N - 2^{n+1})$	72	392	1,800
2-D Mesh	$2(N - 2^n)$	24	112	480
2-D Torus	$4(N - 2^n)$	48	224	960

is $N = 2^n \times 2^n$, $l_{ht}^i = 2^{i+1}$ and $r_{ht}^i = N/4^i$, where $1 \le i \le n$. Therefore, Equation 3.5 can be transformed as follows.

$$L_{2D,ht} = \sum_{i=1}^{n} l_{ht}^i \cdot r_{ht}^i = \sum_{i=1}^{n} 2^{i+1} \cdot \frac{N}{4^i} = 2(N - 2^n) \qquad (3.6)$$

Similarly, the total unit-length of links in a Fat Tree (2,4,1) network, $L_{2D,ft1}$, is

$$L_{2D,ft1} = \sum_{i=1}^{n} l_{ft1}^i \cdot r_{ft1}^i = \sum_{i=1}^{n} 2^{i+1} \cdot \frac{N}{2^{i+1}} = nN. \qquad (3.7)$$

A Fat Tree (2,4,2) has double the number of routers in the Fat Tree (2,4,1); so $L_{2D,ft2} = 2L_{2D,ft1}$. A Fat H-Tree has two folded H-Tree networks, in which each link, except for the links connecting to the top-rank router, requires twice the wire resources of an ordinary H-Tree. By folding the H-Tree, only the top-rank router and its four child routers can be placed inside a 1-unit × 1-unit grid square. Thus, the total unit-length of links in a Fat H-Tree, $L_{2D,fht}$, can be expressed as follows.

$$l_{fht}^i = \begin{cases} 2l_{ht}^i & 1 \le i \le n-1 \\ 4 & i = n \end{cases} \qquad (3.8)$$

$$L_{2D,fht} = \sum_{i=1}^{n} l_{fht}^i \cdot r_{fht}^i = l_{fht}^n \cdot r_{fht}^n + \sum_{i=1}^{n-1} 2^{i+2} \cdot \frac{2N}{4^i} = 8 + 8(N - 2^{n+1}) \qquad (3.9)$$

The total unit-lengths of the 2-D layouts mentioned above are summarized in Table 3.4. As for the mesh and torus, we ignore the links between the core and router, which will increase the total unit-length, for the sake of simplicity. Although a Fat H-Tree uses slightly more wire resources compared to the Fat Tree (2,4,2) in 16- and 64-core networks, the impact on the chip design is considered modest, because enormous wire resources are available in an NoC, thanks to current CMOS technology, which has eight or more metal layers.

The longest link in a network affects the wire delay and the number of repeater buffers inserted in the wires. As for the Fat H-Tree, each link, except for the links connecting to the top-rank router, requires twice the wire length of a same-sized H-Tree, while the length of the top-rank link is 1-unit because of its folded layout, as mentioned above. Thus, the longest link length in a Fat H-Tree is the same as those in the H-Tree and the Fat Trees.

3.4 Routing

The packet routing decides routing paths for a given source and destination nodes. It is crucial to make the best use of topological bandwidth of a given network.

3.4.1 *Deadlocks and Livelocks*

The routing should resolve the deadlock problem. Figure 3.13 shows a situation where every packet is blocked by each other, and they cannot be permanently forwarded. Such a cyclic dependency is called a "deadlock". Once a deadlock occurs, at least a single packet within a network must be killed and resent. To avoid deadlocks, "deadlock-free routing" algorithms that never introduce deadlocks on paths have been widely researched.

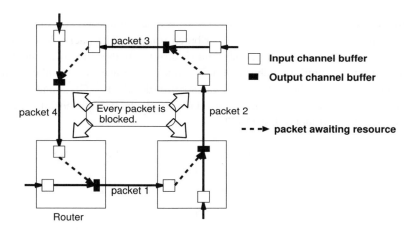

Fig. 3.13   Deadlocks of packets

Besides the deadlock-free property, routing must have livelock-free property in order to avoid packet transfers that never reach the destination. Packets would not arrive at their destinations if they continue to use non-minimal paths that go away from destination nodes. If they were the case, they would be permanently forwarded within NoCs. This situation is called "livelock".

The deadlock- and livelock-free properties are not strictly required to routing algorithms in the case of traditional LAN and WAN. This is because Ethernet usually employs a spanning tree protocol that limits the topology to that of a tree whose structure does not cause deadlocks of paths; moreover, the Internet protocol allows packets to have time-to-live field that limits the maximum number of transfers. Thus, the NoC routing cannot simply borrow the techniques used by commodity LANs and WANs, and new research fields have grown up dedicated to NoCs similar to those in parallel computers.

3.4.2 Routing Algorithm

Since the set of minimal paths on tree-based topologies such as H-Tree is naturally deadlock-free, we omit discussion of routing algorithms on tree-based topologies.

Here, we focus on the routing algorithms on mesh and torus topologies.

Dimension-Order Routing A simple and popular deterministic routing is dimension-order routing (DOR), which uses y-dimension channels after using x-dimension channels in 2-D tori and meshes. Dimension-order routing uniformly distributes minimal paths between all pairs of nodes. It is sometimes referred as xy routing or e-cube routing. Dimension-order routing is usually implemented with simple combinational logic on each router; thus, routing tables that require register files or memory cells for storing routing paths are not used.

Turn Model Assume that a packet moves to its destination in a 2-D mesh topology and routing decisions are implemented as a distributed routing. The packet has two choices in each hop. That is, it decides to go straight or turn to another dimension at every hop along the routing path until it reaches its final destination. However, several combinations of turns can introduce cyclic dependencies that cause deadlocks. Glass and Ni analyzed special combinations of turns that never introduce deadlocks [Glass (1992)]. Their model is referred to as Turn-Model [Glass (1992)].

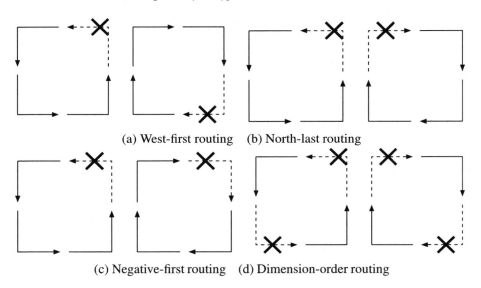

(a) West-first routing   (b) North-last routing

(c) Negative-first routing   (d) Dimension-order routing

Fig. 3.14   Prohibited turn sets of three routing algorithms in Turn-Model

In addition, for 2-D mesh topology, they proposed three deadlock-free routing algorithms by restricting the minimum sets of prohibited turns that may cause deadlocks. These routing algorithms are called West-first routing, North-last routing, and Negative-last routing [Glass (1992)]. Figure 3.14 shows the minimum sets of prohibited turns in these routing algorithms. In West-first routing, for example, turns from the north or south to the west are prohibited. The authors of [Glass (1992)] proved that deadlock freedom is guaranteed if these two prohibited turns are not used in 2-D mesh.

Turn-model view can also demonstrate the deadlock-freedom of dimension-order routing. Figure 3.14.(d) shows a set of prohibited turns in dimension-order routing. The prohibited turn-set in dimension-order routing includes that of West-first routing. That is, the restriction of dimension-order routing is a superset of West-first routing's one. Dimension-order routing additionally prohibits two turns, which are not prohibited in West-first routing; thus the routing diversity of West-first routing is better than that of dimension-order routing. However, a routing algorithm that has high routing diversity does not always outperform one with lower diversity. In fact, West-first routing algorithm can select various routing paths, and it may select a routing path set with a very poor traffic distribution that has hot spots. On the other hand, dimension-order routing can always uniformly distribute the traffic and achieve good performance in the cases of uniform traffic, even though its path diversity is poor.

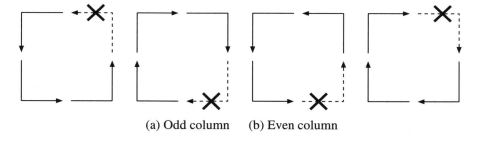

(a) Odd column   (b) Even column

Fig. 3.15   Prohibited turn set in Odd-Even Turn-Model

Chiu extended the turn model into one called Odd-Even Turn-Model [Chiu (2000)], in which nodes in odd columns and even columns prohibit different sets of turns. Figure 3.15.(a) shows a prohibited turn-set for nodes in odd columns, while Figure 3.15.(b) shows one for nodes in even columns. As reported in [Chiu (2000)], the routing diversity of Odd-Even Turn-Model is better than those of the original turn models proposed by Glass and Ni. Thus, Odd-Even Turn-Model has an advantage over the original ones, especially in networks with faulty links that require a higher path diversity to avoid them.

Turn models can guarantee deadlock-freedom in 2-D mesh, but they cannot remove deadlocks in rings and tori that have wrap-around channels in which cyclic dependencies can be formed. A virtual channel mechanism is typically used to cut such cyclic dependencies. That is, packets are first transferred using virtual-channel number *zero* in tori, and the virtual-channel number is then increased when the packet crosses the wrap-around channels.

Duato's Protocol   Duato gave a general theorem defining a criterion for deadlock free-dom and used the theorem to develop a fully adaptive, profitable, progressive protocol [Du-ato (1995)], called Duato's protocol or *-channel. Since the theorem states that by separat-ing virtual channels on a link into escape and adaptive partitions, a fully adaptive routing can be performed and yet be deadlock-free. This is not restricted to a particular topology or routing algorithm. Cyclic dependencies between channels are allowed, provided that there exists a connected channel subset free of cyclic dependencies.

A simple description of Duato's protocol is as follows.

a. Provide an escape path in which every packet can always find a path toward its destina-tion whose channels are not involved in cyclic dependencies.
b. Guarantee that every packet can be sent to any destination node using an escape path and the other path on which cyclic dependency is broken by the escape path (fully adaptive path).

By selecting these two minimal routes (escape path and fully adaptive path) adaptively, deadlocks can be prevented.

Duato's protocol can be extended to arbitrary topologies by allowing more routing restric-tions, and non-minimal paths [Silla (2000)].

Power-, Reliable-, or Traffic-Aware Routing   In addition to the above traditional rout-ings, there are a large number of studies on routings for improving the power consumption, reliability, or performance optimization to the target traffic patterns.

Fault-tolerant routing algorithms whose paths avoid hard faulty links or faulty routers were thus proposed [Flich (2007); Murali (2006)]. These routing tables tend to be complex, or large in order to employ flexible routing paths compared with those of dimension-order routing that has the strong regularity. Thus, the advanced techniques that decrease the table size have been proposed [Koibuchi (2006); Flich (2007); Bolotin (2007)].

Energy-aware routing strategy tried to minimize the energy by improving routing algo-rithm [Kao (2005)]. Dynamic power consumption is proportional to the square of the supply voltage; because a peak performance is not always required during the whole exe-cution time, adjusting the frequency and supply voltage so as to at least achieve the required performance can reduce the dynamic power. In [Stine (2004)], the network link voltage is scaled down by an adaptive routing to distribute the traffic load.

Another routing approach used dynamic traffic information for improving the perfor-mance [Matsutani (2005, 2006)]. DyAD routing forwards packets using deterministic rout-ing at the low traffic congestion, while it forwards them using adaptive routing at the high traffic congestion [Hu (2004)]. DyAD strategy achieves low latency using deterministic routing when a network is not congested, and high throughput by using adaptive routing. DyXY adaptive routing improves throughput by using the congestion information [Li (2006)].

3.5 Summary

In this chapter, we described the router architecture, network topology, and deadlock-free routing design for on-chip interconnection networks of multi- and many-core systems.

In Section 3.2, we explained the switching techniques and conventional router architecture. We surveyed various low-latency router designs including the speculative router, look-ahead router, and bypassing router. We introduced some low-power techniques including the power gating and DVFS. We also discussed the reliability issues of on-chip routers. In Section 3.3, we introduced various network topologies which are applicable to NoCs. These topologies were analyzed in terms of the ideal throughput, average hop count, number of routers, and wire length in order to reveal their pros and cons. Section 3.4 discussed the packet routing. It first illustrated the deadlock and livelock situations. Then it introduced some deadlock-free routing algorithms which can be used in NoCs.

Chapter 4

Parallelizing Compiler for High Performance Computing

To overcome challenges from high power densities and thermal hot spots in microprocessors, multicore computing platforms have emerged as the ubiquitous computing platform from servers to embedded systems. But, providing multiple cores does not directly translate into increased performance for most applications. The burden is placed on software developers to find and exploit coarse-grain parallelism to effectively make use of the abundance of computing resources provided by the systems. With the rise of multi-core systems and many-core processors, concurrency becomes a major issue in the daily life of a programmer. Thus, compiler and software development tools will be critical to help programmers create high performance software This chapter covers software issues of a so called parallelizing queue compiler targeted for future single and multicore embedded systems.

4.1 Instruction Level Parallelism

Instruction level parallelism (ILP) is the key to improve the performance of modern architectures. ILP allows the instructions of a sequential program to be executed in parallel on multiple data paths and functional units. Data and control independent instructions determine the groups of instructions that can be issued together while keeping the program correctness [Muchnick (1997)]. A good scheduling is crucial to achieve high performance. An effective scheduling for the exploitation of ILP depends greatly on two factors: the processor features, and the compiler techniques. In superscalar processors, the compiler exposes ILP by rearranging instructions. However, the final schedule is decided at runtime by the hardware [Hennessy (1990)]. In VLIW machines, the scheduling is decided at compile-time by aggressive static scheduling techniques [Allen (2002); Muchnick (1997)]. Sophisticated compiler optimizations have been developed to expose high amounts of ILP in loop regions [Wolfe (1996)] where many scientific and multimedia programs spend most of their execution time. The purpose of some loop transformations such as loop unrolling is to enlarge basic blocks by combining instructions called in multiple iterations to a single iteration. A popular loop scheduling technique is modulo scheduling [Rau (1994); Lam (1988)] where the iterations of a loop are parallelized in such a way that a new iteration initiates before the previous iteration has completed execution. These static scheduling algorithms improve greatly the performance of the applications at the cost of increasing the register pressure [LOca (1998)]. When the schedule requires more registers than those

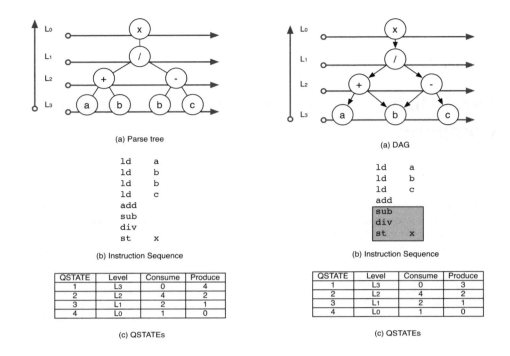

(a) Parse tree

```
ld    a
ld    b
ld    b
ld    c
add
sub
div
st    x
```

(b) Instruction Sequence

QSTATE	Level	Consume	Produce
1	L3	0	4
2	L2	4	2
3	L1	2	1
4	L0	1	0

(c) QSTATEs

(a) DAG

```
ld    a
ld    b
ld    c
add
sub
div
st    x
```

(b) Instruction Sequence

QSTATE	Level	Consume	Produce
1	L3	0	3
2	L2	4	2
3	L1	2	1
4	L0	1	0

(c) QSTATEs

Fig. 4.1 Instruction sequence generation from the parse tree of expression $x = \frac{a+b}{b-c}$.

Fig. 4.2 Instruction sequence generation from DAG of expression $x = \frac{a+b}{b-c}$.

available in the processor, the compiler must insert spill code to fit the application in the available number of architected registers [Printer (1993)]. Many high performance architectures born in the last decade [Sparc (1992); Kane (1992); Kessler (1999)] were designed on the assumption that applications could not make effective use of more than 32 registers [Mahlke (1992)]. Recent studies have shown that the register requirements for the same kind of applications using the current compiler technology demands more than 64 registers [Postiff (2000)]. High ILP register requirements has direct impact in the processor performance as a large number of registers need to be accessed concurrently. The number of ports to access the register file affect the access time and the power consumption. In order to maintain clock speed and low power consumption, high performance embedded, and digital signal processors have implemented partitioned register banks [Janssen (1995)] instead of a large monolithic register file. Several software solutions for the compiler have been proposed to reduce the register requirements of modulo schedules [Salamea (2004)], and other studies have focused on the compilation issues for partitioned register files [Jang (1998); Huang (2001)]. A hardware/compiler technique to alleviate register pressure is to provide more registers than allowed by the instruction encoding. In [fernandes (1997); Tyson (2001)] the usage of queue register files has been proposed to store the live variables in a software pipelined loop schedule while minimizing the pressure on the architected registers. The work in [Ravindran (2205)] proposes the use of register windows to give the illusion of a large register file without affecting the instruction set bits.

An alternative to hide the registers from the instruction set encoding is by using a queue machine. A queue machine uses a first-in first-out structure, called the operand queue, as the intermediate storage location for computations. Instructions read and write the operand queue implicitly. Not having explicit operands in the instructions make instructions short improving code density. Also false dependencies disappears from programs eliminating the need for register renaming logic that reduces circuitry and improves power consumption [Kucuk (2003)]. Queue computers have been studied in several works. Bruno [Preiss (1985)] was the first who investigated the possibility of evaluating expression trees and highlighted the problems of evaluating directed acyclic graphs (DAG) in an abstract queue machine. In [Okamoto (1999)], Okamoto proposed the hardware design of a superscalar queue machine. Schmit et. al [Schmit (2002)] use a queue machine as the execution layer for reconfigurable hardware. They transform the program's data flow graph (DFG) into a spatial representation that can be executed in a simple queue machine. This transformation inserts extra special instructions to guarantee correct execution by allowing every variable to be produced and consumed only once. Their experiments show that the execution of programs in their queue machine have the potential of exploiting high levels of parallelism while keeping code size less than a RISC instruction set. On our previous work [Ben-Abdallah (2006,?); Sowa (2005)], we designed a 32-bit QueueCore processor with a 16-bit instruction set format. Our approach is to allow variables to be produced only once but can be consumed multiple times. We sacrifice some bits in the instruction set for an offset reference to indicate the relative location of a variable to be reused. The goal is to allow DAGs to be executed without transformations that increase the instruction count while keeping reduced instructions that generate dense programs.

Ideas about compiling for queue machines have been discussed in the previous work in an abstract way. Some problems have been clearly identified but no algorithms have been proposed. Before, we explored the possibility of using a retargettable code generator for register machines to map register code into the queue computation model [Canedo (2006)]. The resulting compiler mapped the operand queue in terms of a large number general purpose registers in the machine description file that is used by the code generator in order to avoid spill code. This approach led to complex algorithms to map register programs into queue programs, excessively long programs, poor parallelism, and poor code quality.

In this paper we present a new code generation scheme implemented in a compiler for the QueueCore processor. Our compiler generates assembly code from C programs. This compiler is, for the best of our knowledge, the first automated code generation tool designed for the queue computation model. The queue compiler exposes *natural* ILP from the input programs to the QueueCore processor. Experimental results show that our compiler can extract more parallelism for the QueueCore than an ILP compiler for a RISC machine, and also generates programs with lower code size.

4.2 Parallel Queue Compiler

The Queue Computation Model (QCM) is the abstract definition of a computer that uses a first-in first-out data structure as the storage space to perform operations. Elements are inserted, or en-queued, through a write pointer named QT that references the rear of the queue. And elements are removed, or dequeued, through a read pointer named QH that references the head of the queue.

The QueueCore is a 32-bit processor with a 16-bit wide producer order QCM instruction set architecture based on the produced order parallel QCM [Sowa (2005)]. The instruction format reserves 8-bit for the opcode and 8-bit for the operand. The operand field is used in binary operations to specify the offset reference value with respect of QH from which the second source operand is dequeued, $QH-N$. Unary operations have the freedom to dequeued their only source operand from $QH-N$. Memory operations use the operand field to represent the offset and base register, or immediate value. For cases when 8-bit is not enough to represent an immediate value or an offset for a memory instruction, a special instruction named "covop" is inserted before the conflicting memory instruction. The "covop" instruction extends the operand field of the following instruction.

QueueCore defines a set of specific purpose registers available to the programmer to be used as the frame pointer register ($fp), stack pointer register ($sp), and return address register ($ra). Frame pointer register serves as base register to access local variables, incoming parameters, and saved registers. Stack pointer register is used as the base address for outgoing parameters to other functions.

4.2.0.1 *Compiling for 1-offset QueueCore Instruction Set*

The instruction sequence to correctly evaluate a given expression is generated from a level-order traversal of the expressions' parse tree [Preiss (1985)]. A level-order traversal visits all the nodes in the parse tree from left to right starting from the deepest level towards the root as shown in Figure 4.1.(a). The generated instruction sequence is shown in Figure 4.1.(b). All nodes in every level are independent from each other and can be processed in parallel. Every node may consume and produce data. For example, a load operation produces one datum and consumes none, a binary operation consumes two data and produces one. A QSTATE is the relationship between all the nodes in a level that can be processed in parallel and the total number of data consumed and produced by the operations in that level. Figure 4.1.(c) shows the production and consumption degrees of the QSTATEs for the sample expression.

Although the instruction sequence from a directed acyclic graph (DAG) is obtained also from a level-order traversal, there are some cases where the basic rules of en-queueing and dequeueing are not enough to guarantee correctness of the program [Preiss (1985)]. Figure 4.2.(a) shows the evaluation of an expression's DAG that leads to incorrect results. In Figure 4.2.(c), notice that at QSTATE 1 there are three operands produced, and at QSTATE 2 the operations consume four operands. The add operation in Figure 4.2.(b) consumes two operands, a, b, and produces one, the result of the addition $a + b$. The sub operation consumes two operands that should be b, c, instead it consumes operands c, $a + b$.

In our previous work [Sowa (2005)] we have proposed a solution for this problem. We give flexibility to the dequeueing rule to get operands from any location in the operand queue. In other words, we allow operands to be consumed multiple times. The desired operand's location is relative to the head of the queue and it is specified in the instruction as an offset reference, QH$-N$. As the en-queueing rule, *production* of data, remains fixed at QT, we name this model the *Producer Order Queue Computation Model*. Figure 4.1 shows the code for this model that solves the problems in Figure 4.2. Notice that add, sub, div instructions have offset references that indicate the place relative to QH where the operands should be taken. The "sub -1, 0" instruction now takes operand b from QH-1, and operand c from QH itself, QH$+0$. We name the code for this model *P-Code*. This nontraditional computation model requires new compiler support to statically determine the value of the offset references.

Correct evaluation of binary instructions whose both source operands are away from QH using QueueCore's one operand instruction set is not possible. To ensure correct evaluation of this case, a special instruction has been implemented in the processor. The dup instruction takes a variable in the operand queue and places a copy in QT. The compiler is responsible of placing dup instructions to guarantee that binary instructions will have their first operand available always at QH, and the second operand may be taken from an arbitrary position in the operand queue by using QueueCore's one operand instruction set. Let the expression $x = -a/(a + a)$ be evaluated using QueueCore's one offset instruction set, its DAG is shown in Figure 4.2.(a). Notice that the level L_3 produces only one operand, a, that is consumed by the following instruction, neg. The add instruction is constrained to take its first source operand directly from QH, and its second operand has freedom to be taken from QH$-N$. For this case, the dup instruction is inserted to make a copy of a available as the first source operand of instruction add as shown with the dashed line in Figure 4.2.(b). Notice that level L_3 in Figure 4.2.(b) produces two data instead of one. The instruction sequence using QueueCore's one offset instruction set is shown in Figure 4.2.(c). This mechanism allows safe evaluation of binary operations in a DAG using one offset instruction set at the cost of the insertion of dup instructions. The QueueCore's instruction set format was decided from our design space exploration [Canedo (2006)]. We found that binary operations that require the insertion of dup instructions are rare in program DAGs. We believe that one operand instruction set is a good design to keep a balance between compact instructions and program requirements.

4.3   Parallel Queue Compiler Frame Work

There are three tasks the queue compiler must do that make it different from traditional compilers for register machines: (1) constrain all instructions to have at most one offset reference, (2) compute offset reference values, and (3) schedule the program expressions in level-order manner. We developed a C compiler for the QueueCore that uses GCC's 4.0.2 front-end and middle-end. The C program is transformed into abstract syntax tree (AST) by the front-end. Then the middle-end converts the ASTs into a language and machine independent format called GIMPLE [Novillo (2004)]. A set of tree transformations and optimizations to remove redundant code and substitute sequences of code with more efficient sequences is optionally available from the GCC's middle-end for this representation. Although these optimizations are available in our compiler, until this point our primary goal was to develop the basic compiler infrastructure for the QueueCore and we have not validated the results and correctness of programs compiled with these optimizations enabled. We wrote a custom back-end that takes GIMPLE intermediate representation and generates assembly code for the QueueCore processor. Figure ?? shows the phases and intermediate representations of the queue compiler infrastructure.

The uniqueness of our compiler is from the 1-offset code generation algorithm implemented as the first and second phases in the back-end. This algorithm transforms the data flow graph to assure that the program can be executed using a one-offset queue instruction set. The algorithm then statically determines the offset values for all instructions by measuring the distance of QH relative position with respect of each instruction. Each offset value is computed once and remains the same until the final assembly code is generated. The third phase of the back-end converts our middle-level intermediate representation into a linear one-operand low level intermediate code, and at the same time, schedules the program in a level-order manner. The linear low level code facilitates the extraction of natural ILP done by the fourth phase. Finally, the fifth phase converts the low level representation of the program into assembly code for the QueueCore. The following subsections describe in detail the phases, the algorithms, and the intermediate representations utilized by our queue compiler to generate assembly code from any C program.

4.3.1   1-offset P-Code Generation Phase

GIMPLE is a three address code intermediate representation used by GCC's middle-end to perform optimizations. Three address code is a popular intermediate representation in compilers that expresses well the instructions for a register machine, but fails to express instructions for the queue computation model. The first task of our back-end is to expand the GIMPLE representation into QTrees. QTrees are ASTs without limitation in the number of operands and operations. GIMPLE's high-level constructs for arrays, pointers, structures, unions, subroutine calls, are expressed in simpler GIMPLE constructs to match the instructions available in a generic queue hardware.

The task of the first phase of our back-end, 1-offset P-Code Generation, is to constrain the binary instructions in the program to have at most one offset reference. This phase detects the cases when dup instructions need to be inserted and it determines the correct place. The code generator takes as input QTrees and generates leveled directed acyclic graphs (LDAGs) as output. A leveled DAG is a data structure that binds the nodes in a DAG to a levels [Heath (1999)]. We chose LDAGs as data structure to model the data dependencies between instructions and QSTATEs. Figure **??** shows the transformations the C program suffers when it is converted to GIMPLE, QTrees, and LDAGs.

The algorithm works in two stages. The first stage converts QTrees to LDAGs augmented with *ghost nodes*. A ghost node is a node without operation that serves as a mark for the algorithm. The second stage takes the augmented LDAGs and remove all ghost nodes by deciding whether a ghost node becomes a dup instruction or is removed.

4.3.1.1 *Augmented LDAG Construction*

QTrees are transformed into LDAGs by a post-order depth-first recursive traversal over the QTree. All nodes are recorded in a look-up table when they first appear, and are created in the corresponding level of the LDAG together with its edge to the parent node. Two restrictions are imposed over the LDAGs for the 1-offset P-Code QCM.

Definition 4.3.1. A level is an ordered list of elements with at least one element.

Definition 4.3.2. The sink of an edge must be always in a deeper or same level than its source.

Definition 4.3.3. An edge to a ghost node spans only one level.

When an operand is found in the look-up table the Definition 4.3.2 must be kept. Line 5 in Algorithm 4.1 is reached when the operand is found in the look-up table and it has a shallower level (closer to the root) than the new level. The function dag_ghost_move_node() moves the operand to the new level, updates the look-up table, converts the old node into a ghost node, and creates an edge from the ghost node to the new created node. The function insert_ghost_same_level() in Line 8 is reached when the level of the operand in the look-up table is the same to the new level. This function creates a new ghost node in the new level, makes an edge from the parent node to the ghost node, and an edge from the ghost node to the element matched in the look-up table. These two functions build LDAGs augmented with ghost nodes that obey Definitions 4.3.2 and 4.3.3.

4.3.1.2 *dup instruction assignment and ghost nodes elimination*

The second and final stage of the 1-offset P-Code generation algorithm takes the augmented LDAG and decides what ghost nodes are assigned to be a dup node or eliminated from the LDAG. The only operations that need a dup instruction are those binary operations whose both operands are away from QH. The augmented LDAG with ghost nodes facilitate the task of identifying those instructions. All binary operations having ghost nodes as their left and right children need to be transformed as follows. The ghost node in the left children is substituted by a dup node, and the ghost node in the right children is eliminated from the LDAG. For those binary operations with only one ghost node as the left or right children, the ghost node is eliminated from the LDAG. Algorithm 4.2 describes the function dup_assignment().

Algorithm 4.1 dag_levelize_ghost (tree t, level)

1: nextlevel $\Leftarrow$ level + 1
2: match $\Leftarrow$ lookup (t)
3: **if** match $\neq$ null **then**
4:      **if** match.level $<$ nextlevel **then**
5:          relink $\Leftarrow$ dag_ghost_move_node (nextlevel, t, match)
6:          **return** relink
7:      **else if** match.level $=$ lookup (t) **then**
8:          relink $\Leftarrow$ insert_ghost_same_level (nextlevel, match)
9:          **return** relink
10:     **else**
11:         **return** match
12:     **end if**
13: **end if**
14: /* Insert the node to a new level or existing one */
15: **if** nextlevel $>$ get_Last_Level() **then**
16:     new $\Leftarrow$ make_new_level (t, nextlevel)
17:     record (new)
18: **else**
19:     new $\Leftarrow$ append_to_level (t, nextlevel)
20:     record (new)
21: **end if**
22: /* Post-Order Depth First Recursion */
23: **if** t is binary operation **then**
24:     lhs $\Leftarrow$ dag_levelize_ghost (t.left, nextlevel)
25:     make_edge (new, lhs)
26:     rhs $\Leftarrow$ dag_levelize_ghost (t.right, nextlevel)
27:     make_edge (new, rhs)
28: **else if** t is unary operation **then**
29:     child $\Leftarrow$ dag_levelize_ghost (t.child, nextlevel)
30:     make_edge (new, child)
31: **end if**
32: **return** new

4.3.2 *Offset Calculation Phase*

Once the LDAGs including dup instructions have been built, the next step is to calculate the offset reference values for the instructions. Following the definition of the producer order QCM, the offset reference value of an instruction represents the distance, in number of queue words, between the position of QH and the operand to be dequeued. The main challenge in the calculation of offset values is to determine the QH relative position with respect of every operation. We define the following properties to facilitate the description

---

Algorithm 4.2 dup_assignment (Node i)

1: **if** isBinary (i) **then**
2:      **if** isGhost (i.left) and isGhost (i.right) **then**
3:          dup_assign_node (i.left)
4:          dag_remove_node (i.right)
5:      **else if** isGhost (i.left) **then**
6:          dag_remove_node (i.left)
7:      **else if** isGhost (i.right) **then**
8:          dag_remove_node (i.right)
9:      **end if**
10:     **return**
11: **end if**

---

of the algorithm to find the position of QH with respect of any node in the LDAG.

Definition 4.3.4. An α-node is the first element of a level.

Definition 4.3.5. The QH position with respect of the α-node of Level-j is always at the α-node of the next level, Level-(j+1).

Definition 4.3.6. A level-order traversal of a LDAG is a walk of all nodes in every level (from the deepest to the root) starting from the α-node.

Definition 4.3.7. The distance between two nodes in a LDAG, $\delta(u,v)$, is the number of nodes found in a level-order traversal between u and v including u.

Definition 4.3.8. A hard edge is a dependence edge between two nodes that spans only one level.

Let p_n be a node for which the QH position must be found. QH relative position with respect of p_n is found after a node in a traversal P_i from p_{n-1} to p_0 (α-node) meets one of two conditions. The first condition is that the node is the α-node, $P_i = p_0$. From Definition 4.3.5, QH position is at α-node of the next level $lev(p) + 1$. The second condition is that P_i is a binary or unary operation and has a hard edge to one of its operands q_m. QH position is given by q_m's following node as a result of a level-order traversal. Notice that q_m's following node can be q_{m+1}, or the α-node of $lev(q_m) + 1$ if q_m is the last node in $lev(q_m)$. The proposed algorithm is described in Algorithm 4.3.

After the QH position with respect of p_n has been found, the only operation to calculate the offset reference value for each of p_n's operands is to measure the distance δ between QH's position and the operand's position as described in Algorithm 4.4.

In brief, for all nodes in a LDAG w, the offset reference values to their operands are calculated by determining the position of QH with respect of every node, and measuring the distance to the operands. Every edge is annotated with its offset reference value.

Algorithm 4.3 qh_pos (LDAG w, Node u)
- 1: $I \Leftarrow$ getLevel (u)
- 2: **for** $i \Leftarrow u.prev$ to $I.\alpha$-node **do**
- 3:   **if** isOperation (i) **then**
- 4:   **if** isHardEdge (i.right) **then**
- 5: $v \Leftarrow$ BFS_nextnode (i.right)
- 6:   **return** v
- 7:   **end if**
- 8:   **if** isHardEdge (i.left) **then**
- 9: $v \Leftarrow$ BFS_nextnode (i.left)
- 10:   **return** v
- 11:   **end if**
- 12:   **end if**
- 13: **end for**
- 14: $L \Leftarrow$ getNextLevel (u)
- 15: $v \Leftarrow L.\alpha$-node
- 16: **return** v

Algorithm 4.4 OpOffset (LDAG w, Node v, Operand r)
- 1: offset $\Leftarrow \delta(\text{qh\_pos}(w,v),r)$
- 2: **return** offset

4.3.3 *Instruction Scheduling Phase*

The instruction scheduling algorithm of our compiler is a variation of basic block scheduling [Muchnick (1997)] where the only difference is that instructions are generated from a
level-order topological order of the LDAGs. The input of the algorithm is an LDAG annotated with offset reference values. For every level in the LDAG, from the deepest level to
the root level, all nodes are traversed from left to right and an equivalent low level intermediate representation instruction is selected for every visited node. Instruction selection
was simplified by having one low level instruction for every high level instruction in the
LDAG representation. The output of the instruction scheduling is a QIR list. QIR is a
single operand low level intermediate representation capable to express the instruction set
of the QueueCore. The only operand is used for memory operations and branch instructions. Offset reference values are encoded as attributes in the QIR instructions. Figure 4.3
shows the QIR list for the LDAG in Figure ??.(d). The QIR includes annotations depicted
in Figure 4.3 with the prefix QMARK_*.

A extra responsibility of this phase is to check code correctness of the 1-offset P-Code
generation algorithm by comparing with zero the value of the offset reference for the first
operand of binary instructions based on the assumption that the 1-offset P-Code generation
algorithm constrains all instructions to have at most one offset reference. For every com-

```
(QMARK_BBSTART (B1))

(QMARK_STMT)
    (QMARK_LEVEL)
        (PUSH_Q (i))
        (PUSH_Q (4))
    (QMARK_LEVEL)
        (LOAD_ADDR_Q (a))
        (MUL_Q)
    (QMARK_LEVEL)
        (ADD_Q)
    (QMARK_LEVEL)
        (SLOAD_Q)
    (QMARK_LEVEL)
        (POP_Q (x))

(QMARK_STMT)
    (QMARK_LEVEL)
        (PUSH_Q (x))
        (PUSH_Q (4))
    (QMARK_LEVEL)
        (LOAD_ADDR_Q (a))
        (MUL_Q)
    (QMARK_LEVEL)
        (ADD_Q)
        (PUSH_Q (7))
    (QMARK_LEVEL)
        (STORE_Q)

(QMARK_STMT)
    (GOTO_Q (L2))
```

Fig. 4.3   QIR code fragment

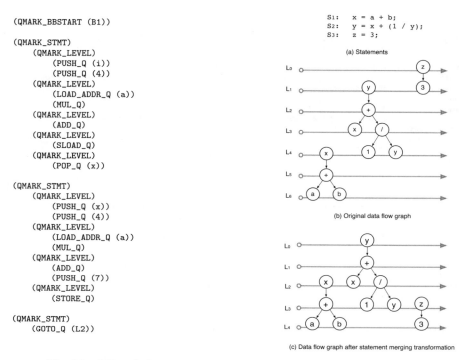

S1:   x = a + b;
S2:   y = x + (1 / y);
S3:   z = 3;

(a) Statements

(b) Original data flow graph

(c) Data flow graph after statement merging transformation

Fig. 4.4   Statement merging transformation

piled function this phase also inserts the QIR instructions for the function's prologue and epilogue.

4.3.4   *Natural Instruction Level Parallelism Extraction: Statement Merging Transformation*

Statement merging transformation reorders the instructions of a sequential program in such a way that all independent instructions from different statements are in the same level an can be executed in parallel following the principle of the QCM. This phase makes a dependence analysis on individual instructions of different statements looking for conflicts in memory locations. Statements are considered the transformation unit. Whenever an instruction is re-ordered, the entire data flow graph of the statement to where it belongs is reordered to keep its original shape. In this way, all offsets computed by the offset calculation phase remain the same, and the data flow graph is not altered. The data dependence analysis looks for two accesses to the same memory location whenever two instructions have the same offset with respect of the base register. Instructions that may alias memory locations are merged safely using a conservative approach to guarantee correctness of the program. Statements with branch instructions and function calls are non-mergeable. Figure 4.4.(a) shows a pro-

```
        ld      ($fp)0  # y                          ld      ($fp)0  # y
        ldil    1                                    ldil    1
        ceq     1       # compare equal              ceq     1       # compare equal
        bt      L1      # branch true                bt      L1      # branch true
LO:                                          LO:
        ld      ($fp)8  # i                           ld      ($fp)8  # i
        ldil    4       # size of a[] element         ldil    4       # size of a[] element
        lda     ($fp)12 # load address of a[0]        lda     ($fp)12 # load address of a[0]
        mul     1       # i * size(a[])               mul     1       # i * size(a[])
        add     1       # address of a[i]             ld      ($fp)4  # x
        lds     0       # load computed address       ldil    4       # size of a[] element
        st      ($fp)4  # x                           add     1       # address of a[i]
        ld      ($fp)4  # x                           lda     ($fp)12 # load address of a[0]
        ldil    4       # size of a[] element         mul     1       # x * size(a[])
        lda     ($fp)12 # load address of a[0]        lds     0       # load computed address
        mul     1       # x * size(a[])               add     1       # address of a[x]
        add     1       # address of a[x]             ldil    7       # rhs constant
        ldil    7       # rhs constant                st      ($fp)4  # x
        sst     0       # st constant in              sst     0       # st constant in
                        # computed address                            # computed address
        j       L2                                   j       L2
L1:                                          L1:
        ld      ($fp)4  # x                           ld      ($fp)4  # x
        ldil    2                                    ldil    2
        mul     1                                    mul     1
        covop   3       # (3<<8) = 768                covop   3       # (3<<8) = 768
        ldil    232     # 768 + 232 = 1000            ldil    232     # 768 + 232 = 1000
        add     1                                    add     1
        st      ($fp)4  # x                           st      ($fp)4  # x
L2:                                          L2:
```

(a) Original QueueCore assembly code   (b) ILP exposed QueueCore Assembly Code

Fig. 4.5   Assembly output for QueueCore processor

gram with three statements S_1, S_2, S_3. The original sequential scheduling of this program is driven by a level-order scheduling as shown in Figure 4.4.(b). When the statement merging transformation is applied to this program a dependency analysis reveals a flow dependency for variable x in S_1, S_2 in levels L_4, L_3. Instructions from S_2 can be moved one level down and the flow dependency on variable x is kept as long the store to memory happens before the load. Statement S_3 is independent from the previous statements, this condition allows S_3 to be pushed to the bottom of the data flow graph. Figure 4.4.(c) shows the DFG for the sample program after the statement merging transformation. For this example, the number of levels in the DFG has been reduced from seven to five.

From the QCM principle, the QueueCore is able to execute the maximum parallelism found in DAGs as no false dependencies occur in the instructions. This transformation merges statements to expose all the available parallelism [Wall (1991)] within basic blocks. With the help of the compiler, QueueCore is able to execute *natural* instruction level parallelism as it appears in the programs. Statement merging is available in the queue compiler as an optimization flag which can be enabled upon user request.

4.3.5 *Assembly Generation Phase*

The last stage of the queue compiler is the assembly code generation for the QueueCore processor. It is done by a one-to-one translation from QIR code to assembly code. The assembly generator is in charge of inserting covop instructions to expand the operand field of those instructions that have operands beyond the limits of the operand field bits. Figure 4.5.(a) shows the generated assembly code and Figure 4.5.(b) shows the assembly code with natural parallelism exposed for the C program in Figure ??.(a). Notice that the original assembly code and the assembly code after statement merging contain exactly the same instructions with the only difference that the order the the instructions change. All instructions have one operand. Depending on the instruction type the only operand has different meaning. The highlighted code fragment in Figure 4.5.(a) shows the assignment of an array element indexed by variable to another variable, in C language "x=a[i]". The first instruction loads the index variable into the queue, its operand specifies the base register and the offset to obtain the memory location of the variable. The operand in the second instruction specifies the immediate value to be loaded, if the value is greater than the instruction bits the assembly phase inserts a covop instruction to extend the immediate value. The operand in the third instruction works is used to compute the effective address of the first element of the array. The next two arithmetic instructions use their operand as the offset reference and help to compute the address of the array element indexed by a variable. For this example, both are binary instructions and take their first operand implicitly from QH and the second operand from QH+1. The lds instruction loads into the queue the value of a computed address taken the operand queue as an offset reference given by its only operand. The last instruction stores the value pointed by QH to memory using base addressing.

To demonstrate the efficiency of our one-offset queue computation model, we developed a C compiler that targets the QueueCore processor. For a set of numerical benchmark programs, we evaluated the characteristics of the resulting queue compiler. We measured the effectiveness of statement merging optimization for improving ILP, we analyzed the quality of the generated code in terms of the distribution of instruction types, and we demonstrate the effectiveness of the queue compiler as a design space exploration tool for our QueueCore by analyzing the maximum offset value required by the chosen numerical benchmarks. To show the potential of our technique for a high-performance processor, we compared the compile-time exposed ILP from our compiler against the ILP exposed by an optimizing compiler for a typical RISC processor. And to highlight the low code size features of our design, we also compare the code size to the embedded versions of two RISC processors. The chosen benchmarks are well known numerical programs: radix-8 fast Fourier transform, livermore loops, whetstone loops, single precision linpack, and quake benchmark. To compare the extracted ILP, we compiled the programs using our queue compiler with statement merging transformation. For the RISC-like processor, we compiled the benchmarks using GCC 4.0.2 with classical and ILP optimizations enabled (-O3) targeting the MIPS I [Kane (1992)] instruction set. The ILP for the QueueCore is measured directly from the DDG in the compiler. The ILP for the MIPS I is measured from

Table 4.1   Lines of C code for each phase of the queue compiler's back-end

Phase	Lines of Code	Description
1-offset P-Code Generation	3000	Code generation algorithm, QTrees and LDAGs infrastructure
Offset Calculation	1500	Algorithm to find the location of QH and distance to each operation
Instruction Scheduling	1500	Level-order scheduling, lowering to QIR, and QIR infrastructure
Statement Merging	1000	Natural ILP exploitation and data dependency analysis
Assembly Generation	1000	Assembly code generation from QIR
Total	**8000**	

Table 4.2   Instruction category percentages for the compiled benchmarks for the QueueCore

Benchmark	Memory	ALU	Move Data	Ctrl. Flow	Covop
fft8g	48.60	47.55	0.32	2.90	0.63
livermore	58.55	33.29	0.20	5.95	4.01
whetstone	58.73	26.73	1.11	13.43	0.00
linpack	48.14	41.59	0.58	8.16	1.52
equake	44.52	43.00	0.56	7.76	3.5

the generated assembly based on the register and memory data dependencies and control flow, assuming no-aliasing information. Code size was measured from the text segment of the compiled programs. MIPS16 [Kissel (1997)] and ARM/Thumb [Goudge (1996)] were chosen for the RISC-like embedded processors. GCC 4.0.2 compiler for MIPS16 and ARM/Thumb architectures was used with full optimizations enabled (-O3) to generate the object files. For the QueueCore, the queue compiler was used with statement merging transformation.

4.3.6 *Results*

The resulting back-end for the QueueCore consists of about 8000 lines of C code. Table 4.1 shows the number of lines for each phase of the back-end.

4.3.6.1 *Queue Compiler Evaluation*

First, we analyze the effect of the statement merging transformation on boosting ILP in our compiler. Figure 4.6 shows the improvement factor of the compiled code with statement merging transformation over the original code without statement merging, both scheduled

Table 4.3   QueueCore's program maximum offset reference value

Benchmark	Maximum Offset
fft8g	29
livermore	154
whetstone	31
linpack	174
equake	89

using the level-order traversal. All benchmarks show an improvement gain ranging from 1.73 to 4.25. The largest ILP improvement is for the fft8g program because it contains very large loop bodies without control flow where the statement merging transformation can work most effectively. Statement merging is a code motion transformation and does not insert or eliminate instructions.

To evaluate the quality of the generated code of our compiler, we organized the QueueCore instructions into five categories: memory, ALU, move data, control flow, and covop. Memory instructions are to load and store to main memory including loading immediate values; ALU includes comparison instructions, type conversions, integer and floating point arithmetic-logic instructions; move data includes all data transfer between special purpose registers; control flow includes conditional and unconditional jumps, and subroutine calls; and covop includes all covop instructions to extend memory accesses and immediate values. Table 4.2 shows the distribution of the instruction categories in percentages for the compiled programs. From the table we can observe that memory operations account for about 50% of the total number of instructions, ALU instructions about 40%, move data less than 1%, control flow less about 8%, and covop about 2%. These results point a place for future improvement of our compiler infrastructure. We believe that classical local and global optimizations [Aho (1986)] may improve the quality of the generated code by reducing the number of memory operations.

The developed queue compiler is a valuable tool for the QueueCore's architecture design space exploration since it gives us the ability to automatically generate assembly code and extract characteristics of the compiled programs that affect the processor's parameters. To emphasize the usage of the queue compiler as a design tool, we measured the maximum offset value required by the compiled benchmarks. Table 4.3 shows the maximum offset value for the given programs. These compiling results show that the eight bits reserved in the QueueCore's instruction format [Ben-Abdallah (2006)] for the offset reference value are enough to satisfy the demands of these numerical calculation programs.

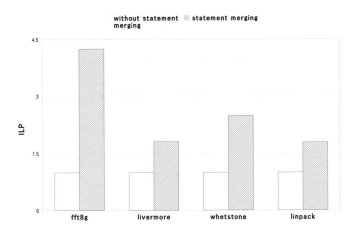

Fig. 4.6   Effect on ILP of statement merging transformation in the queue compiler

4.3.6.2 *Comparison of Generated QueueCore Code with Optimized RISC Code*

The graph in Figure 4.7 compares the ILP improvement of the queue compiler over the optimizing compiler for MIPS processor. For all the analyzed programs, the queue compiler exposed more natural parallelism to the QueueCore than the optimizing compiler for the RISC machine. The improvement of parallelism comes from the natural parallelism found in the level-order scheduled data flow graph with merged statements. QueueCore's instruction set benefits from this transformation and scheduling as no register names are present in the instruction format. The RISC code, on the other hand, is limited by the architected registers. It depends on the good judgement of the compiler to make effective use of the registers to extract as much parallelism as possible, and whenever the register pressure exceeds the limit then spill registers to memory. The loop bodies in livermore, whetstone, linpack, and equake benchmarks consist of one or few instructions with many operands and operations. The improvement of our technique in these programs comes mainly from the level-order scheduling of these "fat" statements since the statement merging has no effect across basic blocks. The greatest improvement on these benchmarks was for the fft8g program which is dominated by manually unrolled loop bodies where the statement merging takes full advantage. In average, our queue compiler is able to extract more parallelism than the optimizing compiler for a RISC machine by a factor of 1.38.

Figure 4.8 shows the normalized code size of the compiled benchmarks for MIPS16, ARM/Thumb and QueueCore using the MIPS I as the baseline. For most of the benchmarks our design achieves denser code than the baseline and the embedded RISC processors. Except for the equake program where the MIPS16 achieved lower code size than the QueueCore. A closer inspection to the object file revealed that the QueueCore program

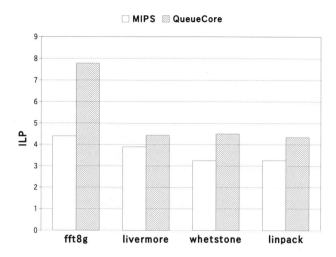

Fig. 4.7   Instruction level parallelism improvement of queue compiler over optimizing compiler for a RISC machine

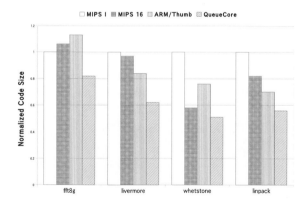

Fig. 4.8   Normalized code size for two embedded RISC processors and QueueCore

has about two times more instructions than the MIPS and MIPS16 code. This is due to the effect of local optimizations such as constant folding, common sub expression elimination, dead code removal, etc. that are applied in the RISC compilers and not in the queue compiler. In average, our design achieves 31 % denser code than MIPS I, 20 % denser code than the embedded MIPS16, and 26 % denser code than the ARM/Thumb processor.

4.4 Conclusion

This chapter presented software issues for design of a parallelizing Queue compiler targeted for single and multicore systems. The design eliminates the register pressure by hiding completely the register file from the instruction set while maintaining a short instruction format with one operand reference. The queue compiler takes advantage of this design and it is capable to expose the maximum natural parallelism available in the data flow graph by means of a statement merging transformation. We evaluated our design by comparing the compile-time extracted parallelism against an optimizing compiler for a traditional RISC machine for a set of numerical benchmarks. Our proposed architecture achieved higher ILP by an average factor of 1.38. Furthermore, the QueueCore processor is also a good candidate for an embedded processor from its reduced instruction set achieving more compact code than embedded RISC-like processors by 26 %.

Chapter 5

Dual-Execution Processor Architecture for Embedded Computing

This chapter introduces practical hardware design issues of a dual-execution processor (DEP) architecture targeted for embedded applications. The architecture is based on a dynamic switching mechanism which allows fast and safe switching between two execution entities at runtime. Several hardware optimization and design techniques are described in the following sections.

5.1 Introduction

Generally, the motivation for the design of a new architecture arises from the technological development, which changed gradually the architecture parameters traditionally used in the computer industry. With these growing changes, the computer architect is faced with answering the question what functionality has to be put on a single chip, giving them an extra performance edge.

The nineties show a tendency towards more complex superscalar processors. These processors which have the ability to initiate multiple instructions during the same clock cycle, are the latest series of architectural innovations aimed at producing ever-faster microprocessors. Because individual instructions are the entities being executed in parallel, super scalar processors exploit what is referred to as instruction level parallelism by issuing multiple instructions per cycle to each functional unit when dependencies allow. In these processors instructions are not necessarily executed in order; an instruction is executed when processor resources are available and data-dependencies are not violated. To execute instructions Out-of-Order data dependencies among instructions must be disabled by the scheduling system during compilation, run time, or both.

Dynamic scheduling detects dependencies in a set of dynamic instruction stream. The most general form of dynamic scheduling, the issue and execution of multiple OoO instructions, can significantly enhance system performance. Instructions with no dependencies are executed if they meet the constraints of the issuing algorithm (discussed in later section). There are other issue difficulties: branch predictions, and fast precise interrupt particularly if a fast response time is desired. Interrupts are precise in a processor with Out-of-Order execution, if after the execution of the interrupt routine, the processor state visible to the operating system and application can be reconstructed to the state a processor would have, had all instructions executed in sequence up to the point of an interrupt. Branch, represents about 15% to 30% of the executed instructions for many applications, decreases the effectiveness of multiple issues to functional units if instructions following an undecided branch cannot be issued. Performance may be improved by enabling speculative executions as predicted path information. If the gains on correct paths out-balance the losses from nullifying execution effects on incorrect paths, performance improves.

Simply stated, achieving higher performance means processing a given program in a smaller amount of time. Therefore, it is imperative that cycle time and total execution time (TET) for a given application be considered when investigating a new architecture. The Dynamic Fast Issue algorithm (discussed later in chapter 3) efficiently addresses the aforementioned problems.

From an other hand, as we enter into an era of continued demand for faster and compatible processors as well as different Internet-network appliances using different processor architectures, it becomes extremely complicated and costly to develop a separate processor for every execution model to satisfies this demand. Internet applications, which are generally "stack based", have complicated the task of processors designers as well and users, who generally seek high execution speed or high performance as defined by the literature. However, recently the term "high performance" is questioned again by many processor designers and computer users. Some consider that high performance means high execution speed or low execution time of some given applications. Other defines "high performance" differently. They consider that processors, which support several executions model, are the favorite candidates for high performance awards, since switching from processor to processor lead to difficulty and waste of time. This is true especially when users have different applications written for different execution model (Stack and RISC model for example). In this case, users are obliged to run these two applications separately on different machines. In conventional machines, this problem was somehow solved by direct software translation techniques. However, these techniques still suffer from slow translation speed. Sun Microsystems proposed its Stack-based Java processor so that Java code can execute directly. According to its designers, the JavaChip-I, for example, is a highly efficient Java execution unit design. It delivers up to 20 times the Java performance of x86 and other general-purpose processor architecture, as well as up to five times the performance obtained by just-in-time (JIT) compilers. It is evident that in tern of reduced TET, the solution is better than the indirect way (translation) or the JIT scheme, but in term of compatibility, the processor still suffers from not being able to execute other codes. The dual-execution mode architecture presented in this chapter is addressing this and other problems as a pure-play architectural paradigm, which will integrate Stack and Queue execution model right into the DEP core.

The Queue execution model (QEM) is analogous to the stack execution model (SEM) in that it has operations in its instructions set, which implicitly reference an operand Queue, just as a stack machine has operations, which implicitly reference an operand Queue.

Each instruction removes the required number of operands from the front of the Queue operand, performs some computations, and stores the result of computation into the Queue of operands at the specified offsets from the head of the Queue. The Queue of operand occupies continuous storage locations. A special register, called the Queue front pointer (QFP), contains the address of the first operand in the operand Queue. Operands are retrieved from the front of the Queue by reading the location indicated by the QFP register. Immediately after retrieving an operand, the QFP is incremented so that it points at the next operand in the Queue. Results are returned to the rear of the operand Queue indicated by the Queue rear pointer (QRP). It will be shown later in that, the instruction sequence of such execution mode is easily generated from an expression parse trees.

In Stack mode, implicitly referenced operands are retrieved from the head of the operand stack and results are returned back onto the head of the stack. However, in our QEM, operands are retrieved from the front of the Queue of operands and results are returned to the rear of the Queue of operands. For example consider a "sub" instruction. In stack execution model, the sub instruction pops two operands from the top of the stack (TOS), computes the difference and pushes the results back onto the top of the stack. However, in Queue execution mode (QEM), the *sub* instruction removes two operands from the front of the Queue (FOQ), computes their difference, and puts the results at the rear of the Queue (ROQ) indicated by the RQP.In the former case, the result of the operation is available at the TOS. In the later case, the result is behind any other operand in the Queue. This will have an enormous potential to effectively exploit pipelined ALU with a simple hardware, which, due to their hardware structure, normal SEM obviously cannot guaranty. Another advantage of the QEM, discussed later, is the reduced program size compared to conventional RISC processor.

5.2 System Architecture

The DEP architecture is a 32-bit processor dual execution mode processor that supports queue (QEM) and stack (SEM) execution modes in a single processor core. The QEM mode uses a first-in-first-out data structure as the underlying control mechanism for the manipulation of operands and results. The QEM is analogous to the stack execution model in that it has operations in its instructions set which implicitly reference an OPQ just as a stack machine has operations that implicitly reference an OPS. Each instruction removes the required number of operands from the front of the OPQ, performs computation, and stores the result of computation into the OPQ at the specified offsets from the tail of the queue. The OPQ occupies continuous storage locations. A special register called queue head (QH) contains the address of the first operand in the OPQ. Operands are retrieved from the front of the queue by reading the location indicated by the QH pointer. Immediately after retrieving an operand, the QH is incremented so that it points to the next operand in the queue. Result is returned to the rear of the OPQ, which is indicated by the queue tail (QT). When switched to stack-based mode, the switching circuitry and the computing unit

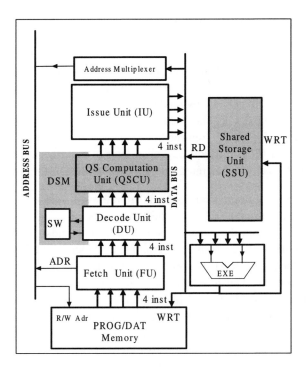

Fig. 5.1   DEP architecture block diagram.

perform the job of executions-mode-switching and calculates the sources and destination operands for corresponding instructions.

In the SEM mode, implicitly referenced operands are retrieved from the top of the OPS and results are returned back into the top of the OPS. For example, consider a sub instruction. In SEM mode, the sub instruction pops two operands from the top of the OPS (TOP), computes the difference and pushes the results back into the top of the stack. However, in QEM mode, the sub instruction removes two operands from the front of the OPQ, computes their difference, and puts the results at the rear of the OPQ. In the former case, the result of the operation is available at the TOP. In the later case, the result is behind any other operand in the queue. This will have an enormous impact to effectively exploit pipelined ALU where normal SEM obviously cannot guarantee. The block diagram of the proposed architecture is shown in Fig. 5.1.

5.2.1   *Pipeline Structure*

The execution pipeline operates in five pipeline stages combined with four pipeline buffers to smooth the flow on instructions through the pipeline. The pipeline stages of DEP processor architecture are following:

Instruction Fetch: Fetches instructions from the program memory and inserts them into a fetch buffer.

Instruction Decode: Decodes instruction's opcodes and operands.

Queue-Stack Computing: The processor's computation stage reads information from the DU and uses them to compute each instruction's sources and destination locations in both queue and stack execution models.

Instructions Issue: The issue stage issues ready instructions to the execution unit.

Execution: Executes issued instructions and sends the results to the Shared Storage Unit or data memory.

5.2.2 *Fetch unit*

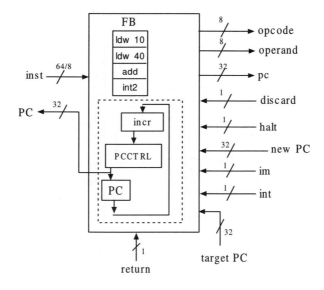

Fig. 5.2   Block diagram of fetch unit.

The fetch mechanism is shown in Fig. 5.2. The fetch unit (FU) has a program counter controller (PCCTRL), an increment register (incr), a program counter register (PC), and a fetch buffer (FB). FU fetches instructions from the program memory. In QEM mode, the FU fetches four instructions per cycle. However, for SEM mode it fetches one instruction per cycle. A special fetch mechanism is used to update the fetch counter according to the mode being processed. A special fetch mechanism is used to update the fetch counter according to the mode being processed. FU controls the fetch mechanism by using the PCCTRL which is controlled by *im*, *halt*, *int*,and *return* signals from different units of processor. The *new PC* and *target PC* forces the PCCTRL for fetch from the their new PC value.

5.2.3 Decode unit

The block diagram of decode unit (DU) is shown in Fig. 5.3.The DU decodes instructions
opcode and operand. DU has 4 decode circuits (DC),1 mode selector register (MS), 1 interrupt vector table (IVT), and 4 covop registers (covop). In DU, we have implemented the
switching mechanism, some part of software interrupt handling mechanism, and the memory address extension mechanism (covop execution).The switching mechanism generates
the instruction mode (im) for each instruction. When the DU detects the software interrupt
instruction it generates the *new PC* and *int* signals. The DU also generates the number of
consumed *CN* and produced *PN* data for each instruction that will be used in the next unit
for calculating the sources and destination addresses. The decode buffer (DB) is a buffer
where the DU inserts the decoded instructions. When DU gets the *discard* signal, it resets
the DB contents. DU detect the halt instruction and generates the *halt* signal for stopping
fetch.

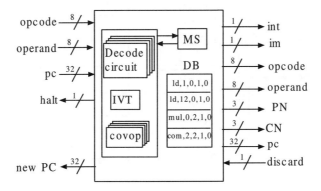

Fig. 5.3   Block diagram of decode unit.

5.2.4 Dynamic switching mechanism

DEP uses a simple dynamic hardware mechanism (DSM) to dynamically switch between
execution modes. The block diagram of the DSM is shown in Fig. 5.4. The DSM system
consists of a switching circuitry (SW) and a dynamic computation unit (QSCU) [Akanda
(2005)]. The QSCU unit calculates the sources and destination addresses for instructions
in both QEM and SEM modes. The DSM detects the instruction mode by decoding the
operand of the "switch" instruction.

After the mode detection, it inserts a mode-bit for all instructions between the current and
the next "switch" instruction. Hence, the same operation can be used for both modes. This
will increase the resources usability and the overall system performance.

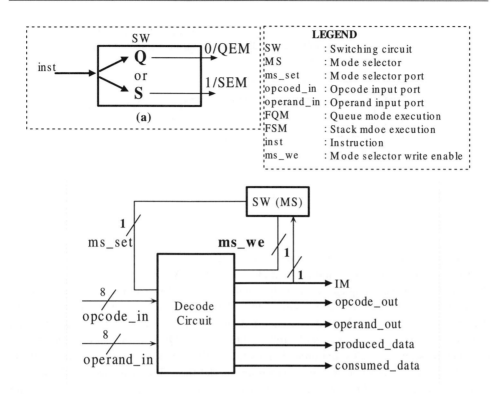

Fig. 5.4   Mode-switching mechanism.

5.2.5   *Calculation of produced and consumed data*

The decode unit calculates the number of PN and CN data by following the Table 5.1. The PN indicates the result of a computation and the CN indicates the operands to be processed. For ALU type instruction, we know that it requires two data for computation and after the computation it makes one result. As a result, the CN for ALU is 2 and PN for ALU is 1. Fig. 5.5 shows the decode mechanism with PN and CN results.

5.2.6   *Queue-Stack Computation unit*

The Queue-stack computation unit (QSCU) is the back bone of DEP processor architecture. The block diagram of QSCU is shown in Fig. 5.6. The queue and stack instruction have their source and destination information as default and here the question is arise why DEP needs the computation unit. The default information is enough for serial execution model. But, our target is to implement the stack computation model on the parallel queue processor architecture. For parallel queue computing the computation unit is extremely necessary for knowing each instruction source and destination addresses and default information is

Table 5.1   PN and CN calculation with instruction in decode stage.
PN means number of produced data and CN means number of consumed data

Instruction	PN	CN
ALU	1	2
Shift	1	1
Set register	0	0
Load	1	0
Store	0	1
Control	0	0

not enough for parallel execution models. So, we have designed a shared computation unit for two different kinds of computation models in QSCU. When program mode is in queue the QSCU behaves like queue computation unit and when the program mode is in stack, the QSCU works like stack computation unit. The QSCU can compute the sources and destination addresses for both queue and stack execution model. We have implemented the *sources-results computing mechanism* in QSCU. The QSCU has two register for calculating the source and destination addresses for parallel queue execution model and these are QH and QT. However, for stack execution model the QT register works like the top of the stack register (TOP). We have designed the *QT/TOP* as shared register for QT and TOP calculation but theoretically it is not shared. The QSCU has a buffer named *queue-stack computation buffer* (QCB). After calculating the source and destination addresses, QSCU sends the instruction (with its source and destination address) to the QCB. When QSCU gets the *discard* signal from execution unit, it resets the QCB contents. The QSCU generates *flag* signal for inform the instruction mode to the next unit (issue).

5.2.7   Sources-Results computing mechanism

We introduce the sources-results computing mechanism for dual execution mode processor architecture. This mechanism is controlled by the instruction mode (IM) which is generated by the DSM. Fig. 5.7 and Fig. 5.8 show the mechanism of sources-results computing that is implemented in QSCU. There are two different circuits (QEM and SEM) controlled by *IM* where the *IM* value is 1 (that means switch is in *on* mode), the QSCU follows the SEM circuit. Otherwise QSCU always follows the QEM circuit.

In QEM execution mode, each instruction needs to know its QH and QT values. The above values are easy to obtain in serial queue execution model, since the QH is always used to fetch instructions from the OPQ and the QT is always used to store the result of the computation into the tail of the OPQ. However, in parallel execution, which is supported by DEP processor, the values for QH and QT are not explicitly determined.

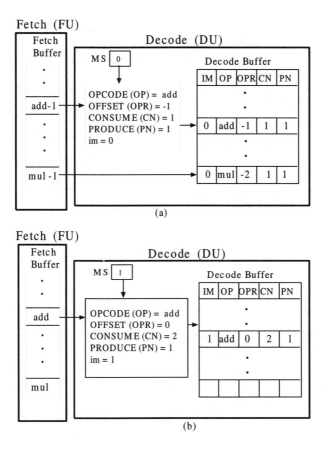

Fig. 5.5   Decode mechanism: (a) decode for queue program and (b) decode for stack program.

Fig. 5.7.(a) shows the hardware mechanism used for calculating *source*1 (first operand), *source*2 (second operand), and destination (result) addresses for current instruction. The computing unit keeps the current value of the QH and QT pointers. In QEM, four instructions arrive at this unit in each cycle. The first instruction uses the current QH ($(QH(n-1))$ in the figure) and QT ($QT(n-1)$) values for *source*1 and destination addresses respectively. As shown in Fig. 5.7.(a), the *source*2 of a given instruction is the first calculated by adding the *source*1 address to the displacement (OFFSET) ($OFFSET(n-1)$).

Fig. 5.8.(a) shows the pointers-update mechanism of current QH and QT values for next instruction. The number of consumed data (CN) field (8-bit) is added to the current QH value (QH0) to find the next QH (NQH) and the number of produced data (PN) field (8-bit) is added to the current QT value (QT0) to find the next QT (NQT). The other three instructions source and destination addresses are calculated similarly.

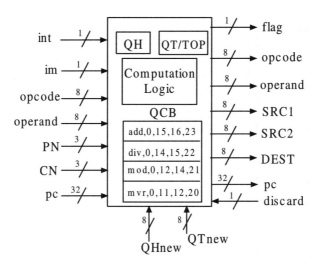

Fig. 5.6   Block diagram of queue-stack computation unit.

In SEM mode, the execution is based on SEM model. The hardware used for calculating *source*1, *source*2, and destination addresses is shown in Fig. 5.7.(b). It is the same hardware used for calculation of operands in QEM mode. The computing unit keeps the current value of the stack pointer (TOP). One instruction arrives to the QSCU unit each cycle. The *source*1 address is popped from the OPS pointed by the current TOP pointer value (TOP). The *source*2 is calculated by subtracting 1 from current TOP pointer value. The number of consumed data (CN) is subtracted from the current TOP value (TOP) and the number of produced data (PN) is added for finding the result address (DEST). Fig. 5.8.(b) shows the hardware mechanism used for calculating the result address for current instruction. The destination address of current instruction points to the next TOP pointer value (NTP), which will be used for next instruction's *source*1 address.

Fig. 5.9.(a) and (b) show two examples of sources and destination addresses calculations for both QEM and SEM execution models respectively. For simplicity, only two instructions "add −1" and "mul −1" are shown in the fetch buffer (FB). The "MS" is the instruction-mode-selector register and is set to 0 which means that the DEP core is in QEM mode. The DU decodes instructions and calculates several fields for each instruction. As shown in the Fig. 5.9.(a), the fields are: IM (instruction mode), OP (opcode), OPR (operand), CN (consumed number), and PN (produced number). The values of IM, OP, OPR, CN, and PN are: 0, add, −1, 1, and 1 for the first instruction "add −1". The same calculation scheme is performed for the following instructions. Using the mechanism shown in Fig. 5.7.(a) and Fig. 5.8.(a), the source1 (QH1), source2 (QH2), and destination (QT0) addresses for each instruction are calculated in the QSCU stage. Fig. 5.9.(b) shows another example illustrating the calculation of TOP, TOP-1, and DEST fields in SEM mode.

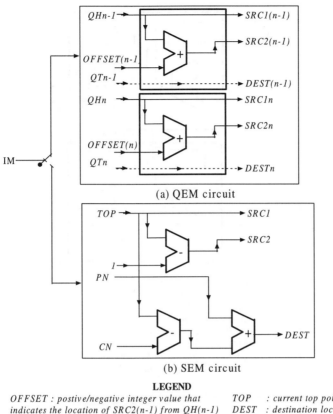

(a) QEM circuit

(b) SEM circuit

LEGEND

OFFSET : postive/negative integer value that	*TOP : current top poiner value*
indicates the location of SRC2(n-1) from QH(n-1)	*DEST : destination location*
QTn : queue tail value of instruction n	*SRC1 : source1 address*
DESTn : destination location of instruction n	*SRC2 : source2 address*
SRC1(n-1) : source address 1 of instruction (n-1)	*IM : instruction mode*
SRC2(n-1) : source address 2 of instruction (n-1)	

Fig. 5.7   Address calculation mechanism for sources and destination.

5.2.8   *Issue unit*

The issue unit (IU) has two issue mechanisms and is controlled by the *flag* signal from QSCU. If the program is in queue mode, IU issues four instructions per cycle. In QEM mode, IU checks the memory and register dependencies. IU also checks the availability of the sources and destination addresses. However, in SEM mode the IU issues only one instruction for sequential execution and in this mode IU does not need to check the dependencies. The Fig. 5.10 shows the block diagram of issue unit. The issue mechanism of DEP looks simple. It is possible for queue programs where always follows the single assignment and there is no false dependency occur in queue and stack storage unit (SSU).

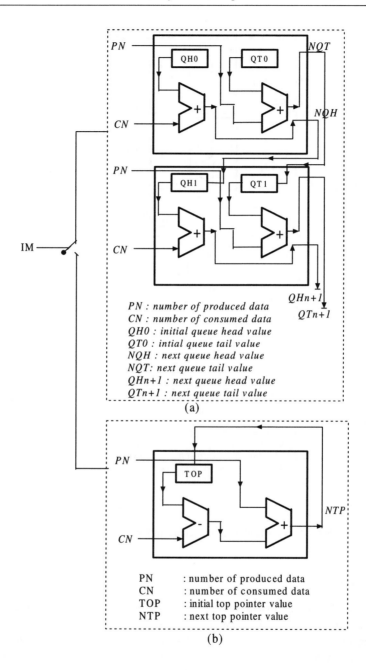

PN : number of produced data
CN : number of consumed data
QH0 : initial queue head value
QT0 : intial queue tail value
NQH : next queue head value
NQT: next queue tail value
QHn+1 : next queue head value
QTn+1 : next queue tail value

(a)

PN   : number of produced data
CN   : number of consumed data
TOP   : initial top pointer value
NTP   : next top pointer value

(b)

Fig. 5.8   Address calculation mechanism for next instruction's source1 and destination.

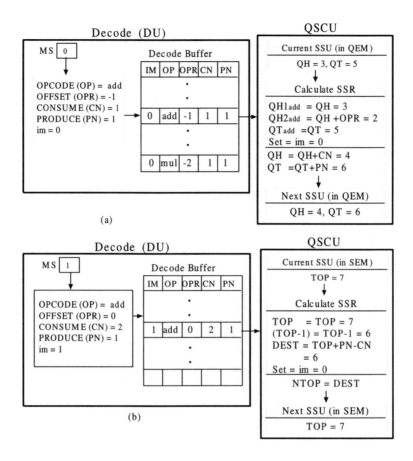

Fig. 5.9   Addresses calculation example: (a) QEM mode and (b) SEM mode.

5.2.9   Execution unit

This is the execution and write back unit of DEP architecture. The block diagram of execution unit (EXE) is shown in Fig. 5.11. The ready instructions come from IU with their appropriate source and destination. EXE executes the ready instructions and sends the result to the shared storage unit (SSU) or memory (DM). If the branch is taken then it sends a *discard* signal to previous all units to stop their activities and also sends the *new PC* to the FU and sends the new qh (*QHnew*) and new qt (*QTnew*) or new top (*QTnew*) pointers value to the QSCU. The block diagram of EXE is shown in Fig. 5.11. Four executing four instructions in parallel in QEM, we have implemented 4 ALU units, 4 Ld/St units, 4 set register units, and 1 control unit in EXE. Now, we do not know the behavior of programs and for getting maximum performance in QEM decided to implement 4 units four each ALU, Ld/St units, and set register unit.

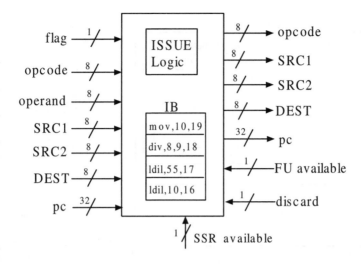

Fig. 5.10   Block diagram of issue unit.

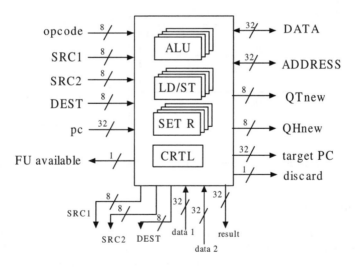

Fig. 5.11   Block diagram of execution unit.

5.2.10   *Shared storage mechanism*

We have implemented the *shared storage mechanism* in the Shared storage unit (SSU). The SSU is an intermediate storage unit for DEP architecture. The SSU has 32-bit 256 shared registers (SSR) and behaves like a conventional register file (RF). However, in QEM, the system organizes the SSR access as a first-in-first-out (FIFO) latches, thus accesses

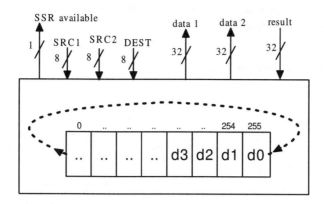

Fig. 5.12   Block diagram of shared storage unit.

concentrate around a small window and the addressing of registers is implicit trough the queue head and tail pointers and in QEM, the system organizes the SSR access as a last-in-first-out (LIFO) latches which addressing the register is implicit through the stack pointer. The shared storage mechanism of SSR is controlled by the QSCU. The block diagram of SSU is shown in Fig. 5.12. The SSU has 4 read and 8 write ports for supporting the maximum performance in QEM mode. The 4 ALU type instruction needs maximum 4 read and 8 write ports of SSU.

5.2.11   *Covop instruction execution mechanism*

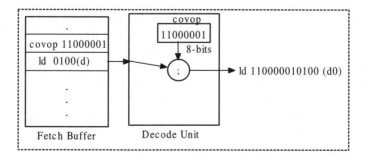

Fig. 5.13   Address extension mechanism.

The *covop* instruction execution mechanism is implemented in DU. Fig. 5.13 shows the main circuit which implements "covop" instruction for extending memory displacement. The mechanism loads the displacement value in the register covop (11000001 in this example). The value in the covop register is then concatenated with the displacement value of the load instruction (0100) and finally generates, the original displacement value(110000010100) for load instruction (ld).

5.2.12 *Interrupt handling mechanism*

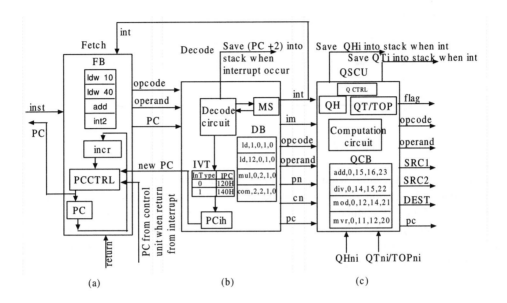

Fig. 5.14   Components used for software interrupt handling mechanism in DEP.

We have implemented the software interrupt handling mechanism in the dual-execution mode processor architecture. The Fig. 5.14 shows the different components used for interrupt handling mechanism. The decode unit (DU) prepares the processor for correct handling of the software interrupt. After an interrupt instruction in the DU is detected, the program counter of the next instruction (PC+2) is saved into the stack. The DU determines the address of the interrupt handler from the interrupt vector table (IVT) and places the value into the PCih. At the same time the DU sends the *int* signal to the fetch unit (FU) and the interrupt (int) signal to the QSCU.

After receiving the *int* signal and the new value for the PC, the fetch unit stops fetching instructions and resets the fetch buffer. Once finished, the FU resumes the fetching starting from the newest value of PC until the return signal coming from the execution unit is detected.

Program Counter Controller: Fig. 5.14.(a) describes the Program Counter Controller (PCC) mechanism. There is a controller (PCCTRL) that selects the PC. The *PCCTRL* is controlled by input signals coming from decode and execution units. Normally, *PCCTRL*

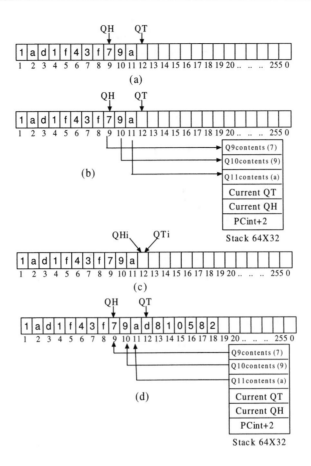

Fig. 5.15 Queue status when interrupt occur and return from interrupt. (a) Queue status before interrupt, (b) when interrupt occur, (c) Queue: ready for interrupt handler, (d) Queue: when return from interrupt.

is selected by normal fetching mode. If there is *int* signal from DU, the PCC is selected by the discard signal and it updates the PC by the new PC. When the PCC gets the return signal with return PC address from the execution unit, it sets the PC as return PC.

Queue-stack status controller: We have implemented a queue-stack status controller (QC-TRL) in QSCU. Fig. 5.14 shows the QCTRL mechanism in QEM. Fig. 5.14.(c) shows the block diagram of QSCU when the interrupt occur and Fig. 5.15 describes the details of queue status before and after the interrupt. Fig. 5.15.(a) shows the present queue contents with present QH and QT pointer addresses. When the QSCU gets the *int* signal from the DU, QSCU saves the current queue status (current QH,QT values, and queue (SSR) contents between QH and QT) into the stack (Fig. 5.15.(b)) and also changes the status of queue. Fig. 5.15.(c) shows the new QH (QHi) and QT (QTi) address for the interrupt han-

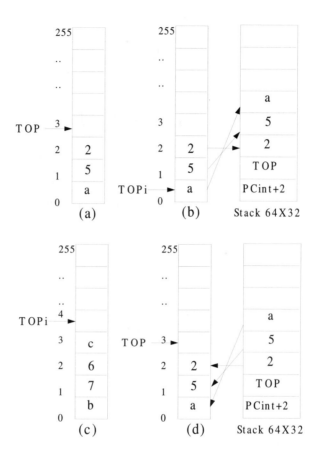

Fig. 5.16 Stack status when interrupt occur and return from interrupt. (a) stack status before interrupt, (b) when interrupt occur, (c) stcak: ready for interrupt handler, (d) stcak: when return from interrupt.

dler. The QSCU will continue the computation by using the QHi and QTi value until it gets the return signal from Execution unit. When QSCU gets the return signal from EU it will restore the QH address, QT address, and queue contents from the stack recovering the normal queue with correct QH and QT pointer address (Fig. 5.15.(d)).

Fig. 5.16 shows the QCTRL mechanism in SEM. In SEM, the QCTRL saves the present TOP value and the stack (SSR) contents into the stack and makes empty the original stack with new TOP pointer(TOPi) (Fig. 5.14.(b)). Then the QSCU will continue the compuation for stack by using (TOPi) until it gets the return signal from EU (Fig. 5.14.(c)). When QSCU gets the return signal form the EU, it will restore the original stack contents with the original TOP address (Fig. 5.14.(d)).

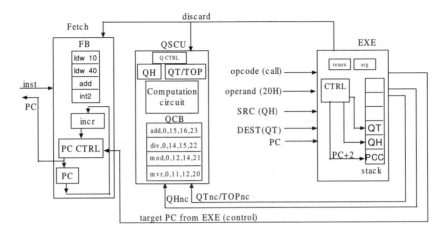

Fig. 5.17   Components used for subroutine call mechanism in DEP.

5.3   Sub-routine call handling mechanism

In DEP architecture, we have also implemented the subroutine call handling mechanism. Fig. 5.17 shows the different components used for subroutine call handling mechanism. We have designed the call execution mechanism in control unit (CTRL) (Fig. 5.17.(c)). After executing the *call* instruction the CTRL creates a *discard* signal and sends to all previous units (FU,DU,QSCU, and IU) for reset their buffers. At the same time CTRL saves the program counter of the next instruction (pc+2) into the stack and sends the new target address for call instruction to the FU. After receiving the *discard* signal and the target address (target PC) form the CTRL, the FU stop fetching and reset the FB. After reset the FB, the FU resumes the fetching starting from the new target PC value until the return signal coming from the execution unit is detected. Here, the program counter controller will do the same job what we described for interrupt handling mechanism. The queue status controller in QSCU resets the QCB and saves the queue status in the stack when it gets the *discard* signal. For saving the return value the CTRL uses a general purpose register (g0) both QEM and SEM mode. In SEM CTRL also uses another general purpose register (g1) for saving the argument. The QCTRL mechanism for subroutine call in QEM is described in Fig. 5.18.

Fig. 5.15.(a) shows the present queue contents with present QH and QT pointer addresses. When the CTRL executes the *call* instruction, it saves the current queue status (current QH,QT values, and queue contents between QH and QT) into the stack (Fig. 5.15.(b)) and also changes the status of queue. The original queue has the argument for the callee program at the position of (QT−1) of the original queue and the QCTRL makes a new queue by following the argument of callee program. So, the new head (QHc) will be pointed by the position of (QT−1) and the new QT (QTc) will be pointed by original QT

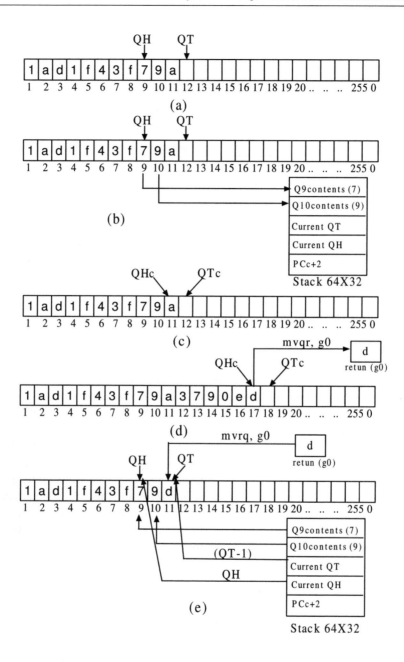

Fig. 5.18   Queue status when subroutine call and return from call. (a) queue status before call, (b) when execute the call, (c) queue: ready for handle callee program, (d) when execute the retun from call (rfc) instruction, and (e) queue: when return from call with retutn result.

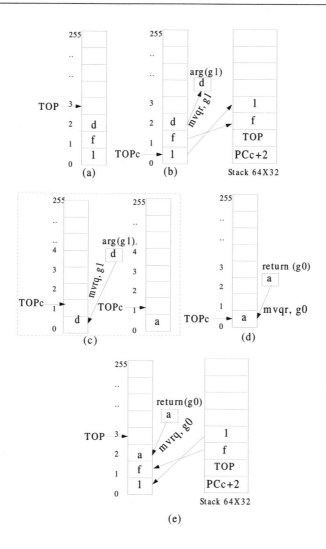

Fig. 5.19   Stack status when subroutine call and return from call.   (a) stack status before call, (b) when execute the call, (c) stack: ready for handle callee program, (d) when execute the retun from call (rfc) instruction, and (e) stack: when return from call with retutn result.

position. Fig. 5.15.(c) shows QHc and QTc address for the callee program. The QSCU will continue the computation by using the QHc and QTc value until it gets the return signal from execution unit.   When the CTRL executes the return form call (rfc) instruction, it stores the return value into a special return register (go) is shown in Fig. 5.18.(d).   This moving data from queue to register (mvqr, g0) is generated by programmer for move the QHc contents to the g0 register. Then QTRL sends the return signal to the QSCU and FU. When QSCU gets the return signal from EU it will restore the QH address, QT address

(QT−1) , and queue contents from the stack recovering the normal queue with correct QH and QT pointer address and the return value from the go register will be move to the QT (Fig. 5.18.(e)). The moving data from register (mvrq, g0) to queue is generated by the programmer for go contents to the QT of the original queue.

Fig. 5.19 shows the QCTRL mechanism for subroutine call in SEM. In SEM the QCTRL first saves the argument for callee into a special *arg* register (g1) and then saves the stack (SSR) contents with the present TOP into the stack of CTRL unit and makes a new stack with new TOP (TOPc). (Fig. 5.19.(b)). For moving the TOP contents to the g1 register the programmer will generate the *mvqr,g1* instruction. Then QCTRL moves the g1 register contents (argument of callee) into new stack (Fig. 5.19.(b)). Then the QSCU will continue the computation by using the TOPc pointer until getting the return signal from EU.

When the CTRL executes the return form call (rfc) instruction, it moves the return value into a special return register (g0) is shown in Fig. 5.19.(d). The programmer will generate the *mvqr,g0* for moving the TOP contents to the g0 register. Then QTRL sends the return signal to the QSCU and FU. When QSCU gets the return signal from EU it will restore the TOP address and stack contents from the stack recovering the normal stack with correct TOP pointer address and the return value from the go register will be move to the TOP (Fig. 5.19.(e)). For moving the g0 contents the programmer generates *mvrq,g0* instruction.

5.4   Hardware design and Evaluation results

To make the DEP design easy to debug, modify, and adapt, we decided to use a high-level description, which was also used by other system designers, such as works in [Maejima (1997); Takahashi (1997)]. We have developed the DEP core in Verilog HDL. After synthesizing the HDL code, the designed processor gives us then the ability to investigate the actual hardware performance and functional correctness. It also gives us the possibility to study the effect of coding style and instruction set architectures over various optimizations. For the DEP core to be useful for these purposes, we identified the following requirements:

(1) High-level description: the format of the DEP description should be easy to understand and modify;
(2) Modular: to add or remove new instructions, only the relevant parts should have to be modified. A monolithic design would make experiments difficult; and
(3) the processor description should be synthesizable to derive actual implementations.

The DEP has been designed with a distributed controller to facilitate debugging and future adaptation for specific application requirements since we target embedded applications. This distributed controller approach replaces a monolithic controller which would be difficult to adapt. The distributed controller is responsible for pipeline flow management and consists of communicating state machines found in each pipeline.

In this design, we have decided to break up the unstructured control unit to small and manageable units. Each unit is described in a separate HDL module. That is, instead of a centralized control unit, the control unit is integrated with the pipeline data path. Thus, each pipeline stage is mainly controlled by its own simple control unit. In this scheme, each distributed state machine corresponds to exactly one pipeline stage and this stage is controlled exclusively by its corresponding state machine. Overall flow control of the DEP processor is implemented by cooperation of the control units in each stage based on communicating state machines. Each pipeline stage is connected to its immediate neighbors and indicates whether it is able to supply or accept new instructions.

In order to estimate the impact of the description style on the target FPGAs efficiency, we have explored logic synthesis for FPGAs. The idea of this experiment was to optimize critical design parts for speed or resource optimizations. In this work, our experiments and the results described are based on the Altera Stratix architecture [Lewis (2002)]. We selected Stratix FPGAs device because it has good trade-offs between routability and logic capacity. In addition it has an internal embedded memory that eliminates the need for external memory module and offers up to 10 Mbits of embedded memory through the TriMatrix TM memory feature. We also used Altera Quartus II professional edition [Altera (2008)] for simulation, placement, and routing. Simulations were also performed with Cadence Verilog-XL tool [Cadence (2008)].

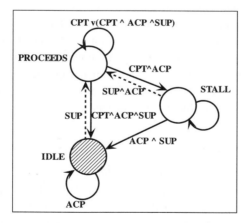

Fig. 5.20   Finite state machine transition for DEP pipeline synchronization

5.4.1   *DEP System pipeline control*

In many conventional processors, the control unit is centralized and controls all central processing core functions. This scheme introduces pipeline stalls, bubbles, etc. However, especially for pipelined architecture, this control unit is one of the most complex part of the design, even for processors with fixed functionality [Lysecky (2005)].

Table 5.2   DEP Processor hardware configuration parameters

Items	Configuration	Description
IW	2 bytes	Instruction width
FW	8 bytes	Fetch width
DW	8 bytes	Decode width
SSR	256	Shared storage register
ALU	4	Arithmetic logical unit
LD/ST	4	Load/Store unit
SET	4	Set unit
BRAN	1	Branch unit
GPR	16	General purpose registers
MEM	2048 word	PROG/DATA memory

Communication with adjacent pipeline stages is performed using two asynchronous signals, *AVAILABLE*, and *PROCEED*. When a stage has finished processing, it asserts the *AVAILABLE* signal to indicate that data is available to the next pipeline stage. The next stage will indicate whether it can proceed these data by using the *PROCEED* signal.

Since all fields, necessary to find what actions are to be taken next, are available in the pipeline stage (for example operation status ready bits and synchronization signals from adjacently stages), computing the next stage is simple. The state transitions of a pipeline stage in the DEP is illustrated in Fig. 6.9. This basic state machine is extended to cover the operational requirements of each stage, by dividing the *PROCEED* state into sub-states as needed. An example is the implementation of the Queue computation stage, where *PROCEED* is divided into sub-states for reading initial addresses values, calculating next addresses values, and addresses fixup (when needed).

5.4.2 *Hardware Design Result*

To simplify the functional verification we developed a front-end tool, which displays the internal state of the processor. The displayed states include: all pipeline stages buffers, SSU and data memory contents. We can easily extend the front-end tool to display several other states at each pipeline stage. For visibility, we only show the state of the SSU and the data memory – the two states are enough for checking the correctness of a simulated benchmark program. This approach allowed us to monitor the program execution as each instruction passes through the pipeline and to identify functional problems by tracing processor state change. We captured the input and output signals changes for several cases.

Table 5.2 shows the hardware configuration parameters of the designed DEP processor.

High-level verification is mainly used to verify functional correctness. Low-level problems such as timing violation cannot be verified directly with high-level verification tools. To ensure correctness of low-level implementation details, interfaces, and timing, we used

Table 5.3   Verilog HDL code size for integrated DEP processor

Description	Modules	Size (Lines)
Fetch unit	fetch.v	192
Decode unit	decode.v	1 059
QS computation unit	qscu.v	327
Issue unit	issue.v	1 038
Execution unit	exe.v	2 630
Shared storage unit	ssu.v	262
Program/data memory	pram.v	50
Top	cpucore.v	1 610
Total		**7 168**

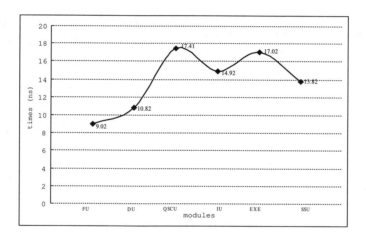

Fig. 5.21   Critical path for different units

gate-level simulation to ensure compliance with design specifications. We specified timing and other constraints using a unified user constraints file with the core modules. We
also used several test benches to verify the correctness of the architecture for both stack
and queue execution models. Table 5.3 shows the design results of the DEP architecture
when synthesized for Stratix FPGA device. The synthesis results are measured by Altera
Quartus II professional edition [Altera (2008)] and form this tool we have got two kinds of
optimization results: speed optimization (SOP) and area optimization (AOP).

The complexities of each module as well as the whole DEP core are given as the number
of logic elements (LEs) for both SOP and AOP optimizations. Table 5.3 shows that the
EXE uses the 31.89 % of total LEs and the SSU uses 51 % of total LEs. In EXE, we
have implemented four ALU, four load/store, and four set register units for executing four

Table 5.4   Synthesis results. LEs means Logic Elements. AOP means Area optimization and SOP means speed optimization

Descriptions	Modules	LE-SOP	LE-AOP
Fetch unit	FU	345	252
Decode unit	DU	1 395	1 269
QS computation unit	QSCU	489	484
Issue unit	IU	1 794	1 588
Execution unit	EXE	7 543	6 213
Shared storage unit	SSU	12 084	8 120
DEP core	DEP	23 650	17 926

Table 5.5   Comparison results between DEP and PQP architecture. Area in LE, Speed in MHz, and Power in mW

Architectures	LE-SOP	LE-AOP	Speed-SOP (MHz)	Speed-AOP (MHz)	Average Power(mW)
DEP	23 650	17 926	64.8	62.31	187.5
PQP	23 065	17 556	71.5	70.1	187.5

instructions in parallel in QEM. For supporting the EXE, we have implemented four write and eight read ports in SSU and SSU has also 32-bit 256 sahred registers. For getting the maximum performance we don't consider about the hardware complexity.

The design was optimized for balanced optimization guided by a properly implemented constraint table. From the prototyping result, the processor consumes 94.389 % of the total logical elements of the target Stratix EP1S25F1020 FPGA device. As a result, the DEP processor successfully fits on a single FPGA device, thereby eliminating the need to perform multi-chip partitioning which results in a loss of resource efficiency.

Fig. 5.21 shows the critical path for each pipeline stage. The evaluation shows that the queue-stack computation stage (QSCU) has the longest delay. This is a largely due to the delays needed for computing the source and destination address for each instruction. The achievable frequency of the DEP core is 64.8 MHz and 62.31 MHz for speed and area optimizations respectively. The performance can be much more improved by using specific layout generation tools and standard libraries. The average power consumption of DEP processor is about 187.5 mW.

Table 5.4 shows the total number LEs for the DEP and the base architecture (PQP). When compared to the PQP core, DEP requires only 2.41 % extra hardware for speed optimization (SOP) and 2.06 % extra hardware for area optimization (AOP). About the speed performance the DEP has loses 12.92 % and 11.88 % speed for SOP and AOP optimizations respectively. The required average power consumption is same both for DEP and PQP ar-

Table 5.6 DEP speed comparisons

Cores	Speed-SOP	Speed-AOP
DEP	64.8	62.31
PQP	71.5	70.1
OpenRisc1200	32.64	32.1
ARM7	25.2	24.5
LEON2	27.5	26.7
MicroBlaze	26.7	26.7
CPU86	26.95	26.77

Table 5.7 DEP power consumption comparisons with various synthesizable CPU cores

Cores	Average Power
DEP	187.5
PQP	187.5
OpenRisc1200	1005
SH7616	250
SH7710	500
SH7720	600
ARM7	22
LEON2	458

chitectures. The comparison results show that with increasing little extra hardware and lose some speed we have implemented the stack computation model on the base PQP architecture that we are expecting from the beginning of DEP research work.

5.4.3 Comparison results

We have already compared the DEP architecture results with PQP aachitecture and both architecture have been designed by us. So, we have compared our DEP processor performance result with other conventional synthesizable processors. Performance of DEP core in terms of speed and power consumption is compared with various synthesized CPU cores as illustrated in Table 5.5 and Table 5.6. The source code for PQP, OpenRisc1200, and CPU86 cores were available. Therefore, we synthesized and evaluated their performance using altera Quartus II professional edition. The other cores data were obtained from corresponding manufacturers performance reports and published work [Mattson (2004); EDN (2008)].

LEON2 is a SPARCV8 compliant 32-bit RISC processor [Gaisler (2004)]. ARM7 is a simple 32-bit RISC core [ARM (1994, 2001)]. The CPU86 core is fully binary/instruction compatible with an 8086/8088 processor [CPU86 (2008)]. The SH2-DSP(SH7616) and SH3-DSP (SH7710, and SH7720) series processor cores are based on a popular Hitachi SuperH (SH) instruction set architecture [IEEE (1997); Hitachi (2007)]. The SH has RISC-type instruction sets and 16×32 bit general purpose registers. All instructions have 16-bits fixed length. The SH2-DSP and SH3-DSP are based on 5 stages pipelined architecture, so basic instructions are executed in one clock cycle pitch. Similar to our processor, the SH also has an internal 32-bit architecture for enhanced data processing ability.

From the results shown in Table 5.5, the DEP core has 49.63 % higher speed than Open-Risc1200 core for both area and speed optimizations. DEP core also has 61.11 %, 57.56 %, 58.53 % and 58.79 % higher speed than ARM7, LEON2, CPU86, and MicroBlaze cores respectively. When compared with PQP processor, DEP has 12.92 % and 11.88 % speed decrease for speed and area optimizations respectively. For power consumption comparison, the DEP core consumes 144.26 % and 33.33 % less power than LEON2 and SH7616 cores respectively. It also consumes less power than SH series and OpenRisc1200 cores. However, DEP core consumes more power than the ARM7 core. This difference comes mainly from the small hardware configuration parameters of ARM7 when compared to our DEP core parameters.

5.5   Conclusions

This chapter presented architecture and design of a dual-execution mode synthesizable processor. The architecture shares a single instruction set and supports both queue and stack execution modes. This is achieved dynamically by execution mode switching and sources-results computing mechanisms. In overall performance, the DEP architecture is expected to increase the processor resources usability, relative to single processor.

Chapter 6

Low Power Embedded Core Architecture

Queue based instruction set architecture processor offers an attractive option in the design of embedded systems. This chapter describes architecture and design results of a low power 32-bit Synthesizable QueueCore (QC-2) with single precision floating point support. This soft core is an efficient architecture which can be integrated in a multicore platform targeted for low power embedded applications.

6.1 Introduction

Since conventional processors are already capable of starting one operation per cycle, reducing CPI further requires starting more than one operation per cycle. To extract ILP, these processors keep many in-flight instructions, use dynamic scheduling, and register renaming [Smith (1995)]. As results, hardware complexity, power dissipation, and resource underutilization in these architectures become critical performance limiting factors [Tiwari (1998)]. There are many efforts in architectures design that address these problems. Thus, computer architects are continuously challenged to bring innovations to design microarchitectures, instruction set, and compilers, which help to keep the balance between performance, complexity and power. Several processors have achieved success as two or four-way Superscalar implementations. However, adding still more functional units is not useful if the rest of the processor is not capable of supplying those functional units with continuous and independent instructions to perform.

A wish to have simple but still fast machine pushed us to look for alternatives. Our research was inspired by several original ideas [Ben-Abdallah (2002); Preiss (1985); fernandes (1997); Heath (1996)], which proposed the usage of queue (first-in-first-out memory) instead of registers (random access memory) as intermediate storage of results. In these architectures, each instruction removes the required amount of data from the head of operand queue (GERG) and then stores the result of computation at the tail of operand queue.

In [Schmit (2002)] it was also argued that a queue machine application can be easily mapped to an appropriate hardware. However, no real hardware was proposed or designed and only theoretical techniques were proposed to virtualize the hardware.

We proposed a produced order parallel Queue processor (PQP) [Sowa (2005); Ben-

Abdallah (2002, 2005, 2006)]. The key ideas of the produced order queue execution model are the operands and results manipulation schemes. The Queue execution model stores intermediate results in a circular queue-register (QREG). A given instruction implicitly reads its first operand from the head (QH) of the QREG, its second operand from a location explicitly addressed with an offset from the first operand location. The computed result is finally written into the QREG at a position pointed by a queue-tail pointer (QT).

The PQP processor has several promising advantages over register-based machines. First, PQP programs have higher instruction level parallelism because they are constructed using breadth-first algorithm [Ben-Abdallah (2002)]. Second, instructions of PQP are shorter because they do not need to specify operands explicitly. That is, data is implicitly taken from the head of operand queue and the result is implicitly written at the tail of the operand queue. This makes instruction lengths shorter and independent from the actual number of physical queue words. Finally, PQP instructions are free from false dependencies. This eliminates the need for register renaming [Ben-Abdallah (2002)].

The QC-2 core implements all hardware features found in PQP core, supports single precision floating-point accelerator, and uses *QCaEXT* scheme - a novel technique used to extend immediate values and memory instruction offsets that were otherwise not representable because of bit-width constraints in the PQP processor. The aim of the *QCaEXT* technique is to achieve code density that is similar to the PQP code with performance similar to architecture set on 32-bit memory. Moreover, the QC-2 core implements new instructions for controlling the QREG unit (discussed later).

We consider the well known prototyping-based emulation approach, that substitutes real time hardware emulation for slow simulator-based execution [Maejima (1997); Takahashi (1997)]. To achieve good synthesis results for FPGAs implementation with sufficient performance, we have created the synthesizable model of the QC-2 processor for the integer and floating subset of the parallel Queue processor instruction set architecture [Sowa (2005); Ben-Abdallah (2002, 2006)]. A prototype implementation is produced by synthesizing the high-level model for the Stratix FPGA device [Lewis (2002); Altera (2008)].

6.2   Produced Order Queue Computing Overview

As mentioned, the produced order queue computing model uses a circular queue-register (also named operand queue) instead of random access registers to store intermediate results. A data is inserted in the QREG in produced order scheme and can be reused. This feature has a profound implication in the areas of parallel execution, program compactness, hardware simplicity and high execution speed [Sowa (2005); Ben-Abdallah (2002)]. This section gives a brief overview about the produced order Queue computation model. We show in Fig. 6.1.(a) a sample data flow graph for the expressions: $e = ab/c$ and $f = ab(c+d)$. Datum is loaded with load instruction (ld), computed with multiply ($*$), add ($+$), and divide ($/$) instructions. The result is stored back in the data memory with store instruction (st).

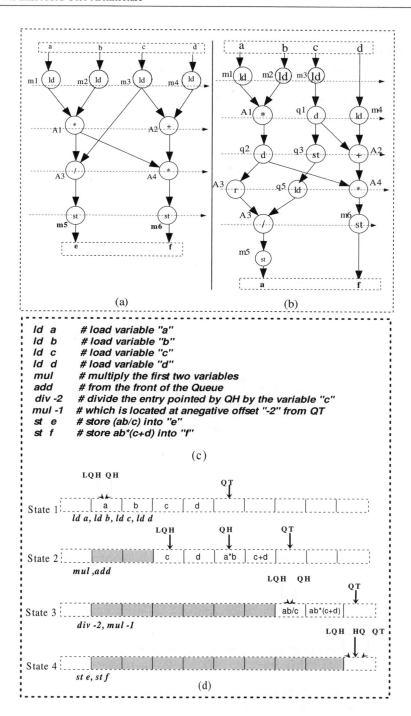

Fig. 6.1   Sample data flow graph and queue-register contents for the expressions: $e = ab/c$ and $f = ab(c+d)$. (a) Original sample program, (b) Translated (augmented) sample program, (c) Generated instructions sequence, (d) Circular queue-register content at each execution state.

In Fig. 6.1.(a), producer and consumer nodes are shown. For example, A2 is a producer node and A4 is a consumer node (A4 is also a producer to m6 node). The instruction sequence for the queue execution model is correctly generated when we traverse the data flow graph (shown in Fig. 6.1.(a)) from left to right and from the highest to the lowest level [Ben-Abdallah (2002)].

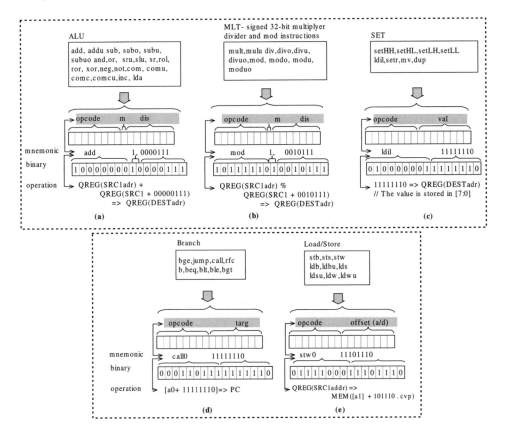

Fig. 6.2   QC-2 instruction format and computing examples: (a) *add* instruction,(b) *mod* instruction, (c) load immediate (*ldil*) instruction, (d) *call* instruction, and (e) store word (*stw*) instruction.

In Fig. 6.1.(b), the *augmented* data flow graph that can be correctly executed in the proposed queue execution model is given. The generated instruction sequence from the *augmented* graph is shown in Fig. 6.1.(c) and the content of the QREG at each execution stage is shown in Fig. 6.1.(d).

A special register, called queue head pointer, points to the first data in the QREG. Another pointer, named queue tail pointer, points to the location of the QREG in which the result data is stored. A live queue head pointer (LQH) is also used to keep used data that could be reused and thus should not be overwritten. These data, which are found between QH and LQH pointers, are called *live-data* (discussed later). The *live-data* entries in the QREG are statically controlled. Two special instructions are used to *stop* or *release* the LQH pointer. Immediately after using the data, the QH is incremented so that it points to the data for the next instruction. QT is also incremented after the result is stored. The four load instructions load in parallel *a*, *b*, *c* and *d* data and place them into the QREG. At this state, QH and LQH point to datum *a* and the QT points to an empty location as shown in Fig. 6.1.(d) (State 1). The fifth and sixth instructions are also executed in parallel. The *mul* refers *a* and *b* then inserts $(a*b)$ into the QREG. QH is incremented by two and the QT is incremented by one. The *add* refers *c* and *d* then inserts $(c+d)$ into the QREG. At this state, the QH, QT and LQH are incremented as shown in Fig. 6.1.(d) (State 2). The seventh instruction $(div - 2)$ divides the data pointed by QH (in this case $(a*b)$) by the data located at -2, negative offset, from QH (in this case c). The QH is incremented and points to $(c+d)$. The eighth instruction multiplies the data pointed by QH (in this case $(c+d)$) with the data located at -1 from QH (in this case $(a*b)$). After parallel execution of these two instructions, the QREG content becomes as shown in Fig. 6.1.(d) State 3. The last two instructions store the result back in the data memory. Since the QREG becomes empty, LQH, QH and QT point to the same empty location (State 4).

6.3   QC-2 Core Architecture

6.3.1   *Instruction Set Design Considerations*

The QC-2 supports a subset of the produced order queue processor instruction set architecture [Ben-Abdallah (2006)]. All instructions are 16-bit wide, allowing simple instructions fetch and decode stages and facilitating pipelining of the processor. The QC-2 integer instruction format is illustrated in Fig. 6.2. Several examples showing the operations of five different instructions are also given in Fig. 6.2.(a)–(e).

In the current version of our implementation, we target the QC-2 core for small applications where our concerns focus on the ability to execute Queue programs on a processor core with small die size and low power consumption characteristics when compared to other conventional 32-bit architectures. However, the short instructions may limit the memory addressing space as only 8-bit are left for offset (6-bit) and base address (2-bit - 00:a0/d0, 01:a1/d1, 10:a2/d2, and 11:a3/d3). To cope with this shortage, QC-2 core implements *QCaEXT* technique, which uses a special "covop" instruction that extends *load* and *store* instructions offsets and also extends immediate values if necessary. The Queue processor compiler [Canedo (2006)] outputs full addresses and full constants and it is the duty of the QC-2 assembler to detect and insert a "covop" instruction whenever an address or a constant exceeds the limit imposed by the instruction's field sizes. Conditional branches are handled in a particular way since the compiler does not handle target addresses, instead it generates target labels. When the assembler detects a target label, it looks if the label has been previously read and fills the instruction with the corresponding value and

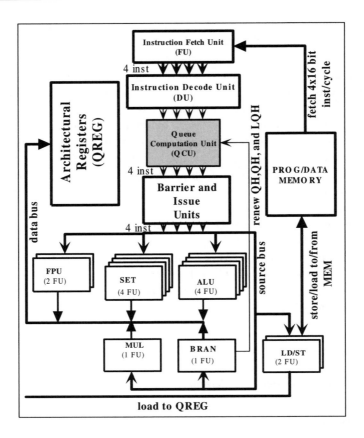

Fig. 6.3   QC-2 architecture block diagram. During RTL description, the core is broken into small and manageable modules using modular approach structure for easy verification, debugging and modification.

"covop" instruction if needed. There is a back-patch pass in the assembler to resolve all missing forward referenced instructions [Canedo (2006)].

6.3.2   Instruction Pipeline Structure

The execution pipeline operates in six stages combined with five pipeline-buffers to smooth the flow of instructions through the pipeline. The QC-2 block diagram is shown in Fig. 6.3. Data dependencies between instructions are automatically handled by hardware interlocks. Below we describe the salient characteristics of the QC-2 core.

(1) *Fetch (FU)*: The instruction pipeline begins with the fetch stage, which delivers four instructions to the decode unit each cycle. This is the same bandwidth as the maximum execution rate of the functional units. At the beginning of each cycle, assuming no pipeline stalls or memory wait states occur, the address pointer hardware (APH) of the fetched

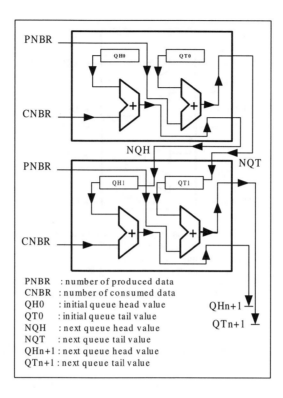

Fig. 6.4   Source 1 (*source1*) address calculation hardware.

instructions issues a new address to the memory system. This address is the previous address plus 8 bytes or the target address of the currently executing flow-control instruction.

(2) *Decode (DU)*: The DU decodes four instructions in parallel during the second phase and writes them into the decode buffer. This stage also calculates the number of consumed (CNBR) and produced (PNBR) data for each instruction [Sowa (2005)]. The CNBR and PNBR are used by the next pipeline stage to calculate the sources (*source1* and *source2*) and destination locations for each instruction. Decoding stops if the queue buffer becomes full or/and a *halt* signal is received from one or more stages following the decode stage.

(3) *Queue computation (QCU)*: Four instructions arrive at the QCU unit each cycle. The QCU calculates the first operand (*source1*) and destination addresses for each instruction. The mechanism used for calculating the *source1* address is given in Fig. 6.4. The QCU unit keeps track of the current value of the QH and QT pointers.

(4) *Barrier:* inserts barrier flags for dependency resolutions.

(5) *Issue (IS):* four instructions are issued for execution each cycle. In this stage, the second operand (*source2*) of a given instruction is first calculated by adding the address *source1* to the displacement that comes with the instruction. The second operand address calculation

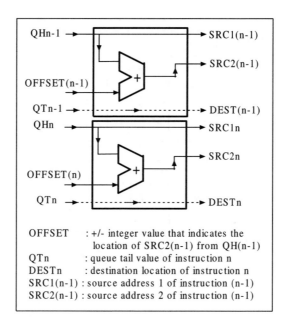

Fig. 6.5   Source 2 (*source2*) address calculation hardware.

is performed in the QCU stage. However, for a balanced pipeline consideration, the *source2* is calculated at the beginning of the IS stage. The hardware mechanism used for calculating the *source2* address is shown in Fig. 6.5 (discussed later).

An instruction is ready to be issued if its data and its corresponding functional unit are available. The processor reads the operands from the QREG in the second half of the IS stage and execution begins in the execution stage (stage 6).

(6) *Execution (EXE)*: The macro data flow execution core consists of 4 integer ALU units, 2 floating-point units, 1 branch unit, 1 multiply unit, 4 set-units, and 2 load/store units.

The load and store units share a 16-entry address window (AW), while the integer units and the branch unit share a 16-entry integer window (IW). The floating-point accelerator (FPA) has its own 16-entries floating point window (FW). The load/store units have their own address generation logic. Stores are executed to memory in-order.

6.3.3   *Dynamic Operands Addresses Calculation*

To execute instructions in parallel, the QC-2 processor must calculate each instruction's operand(s) and destination addresses dynamically. As a result, the "static" Queue data structure (compiler point of view) is regarded dynamically as a circular queue-register structure. Fig. 6.4 and Fig. 6.5 show block diagrams of the hardware used for calculating *source1*, *destination* and *source2* respectively. To calculate the *source1* address of a

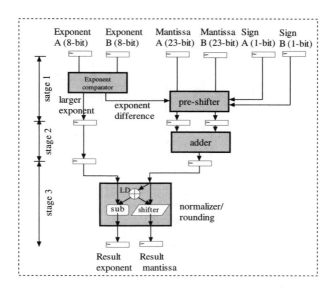

Fig. 6.6   QC-2's FPA Hardware: Adder Circuit

given instruction, the number of consumed data (CNBR) field is added to the current queue head value (QH_n). The destination address on the next instruction ($INST_{n+1}$) is calculated by adding the PNBR field (8-bit) to the current queue tail value (QT_n). Notice that the calculation is performed sequentially. Each QREG entry is written exactly once and it is busy until it is written. If a subsequent instruction needs its value, that instruction must wait until requested data is written. After a given entry in the QREG is written, the corresponding data in the above entry is ready and its ready bit (RDB) is set.

6.3.4   QC-2 FPA Organization

The QC-2 floating-point accelerator is a pipelined structure and implements a subset of the IEEE-754 single precision floating-point standard [IEEE (1985, 1981)]. The FPA consists of a floating-point ALU (FALU), floating-point multiplier (FMUL), and floating point divider (FDIV). The FALU, FMUL, FDIV and the floating-point queue-register employ 32-wide data paths. Most FPA operations are completed within three execution cycles. The FPA's execution pipelines are simple in design for high speed that the QC-2 core requires. All frequently used operations are directly implemented in the hardware. The FPA unit supports the four rounding modes specified in the IEEE 754 floating point standard: round toward-to-nearest-even, round towards positive infinity, round towards negative infinity, and round towards zero.

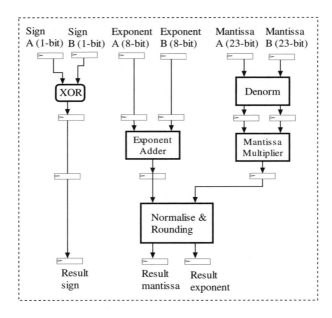

Fig. 6.7   QC-2's FPA Hardware: Multiplier Circuit

6.3.4.1 *Floating Point ALU Implementation*

The FALU does floating-point addition, subtraction, compare and conversion operations. Its first stage subtracts the operands exponents (for comparison), selects the larger operand, and aligns the smaller mantissa. The second stage adds or subtracts the mantissas depending on the operation and the signs of the operands. The result of this operation may overflow by a maximum of 1-bit position. Logic embedded in the mantissa adder is used to detect this case, allowing 1-bit normalization of the result on the fly. The exponent data path computes (E+1). If the 1-bit overflow occurred, (E+1) is chosen as the exponent of stage 3; otherwise, E is chosen. The third stage performs either rounding or normalization because these operations are not required at the same time. This may also result in a 1-bit overflow. Mantissa and exponent corrections, if needed, are implemented exactly in this stage, using instantiations of the mantissa adder and exponent blocks. The area efficient FALU hardware is shown in Fig. 6.6. The exponents of the two inputs (Exponent A and Exponent B) are fed into the exponent comparator, which is implemented with a subtracter and a multiplexer. In the pre-shifter, a new mantissa is created by right shifting the mantissa corresponding to the smaller exponent by the difference of the exponents so that the resulting two mantissas are aligned and can be added. The size of the preshifter is about $m*log(m)LUTs$, where m is the bit-width of the mantissa. If the mantissa adder generates a carry output, the resulting mantissa is shifted one bit to the right and the exponent is increased by one. The normalizer transforms the mantissa and exponent into normalized format. It first uses a leading-one detector (LD) circuit to locate the position of the most significant one in the mantissa. Based on the position of the LD, the resulting mantissa is left shifted by an amount subsequently deducted from the exponent. If there is an exponent overflow (during normalization), the result is saturated in the direction of overflow and the overflow flag is set. Underflows are handled by setting the result to zero and setting an underflow flag. We have to notice that the LD anticipator can be also predicted directly from the input to the adder. This determination of the leading digit position is performed in parallel with the addition step so as to enable the normalization shift to start as soon as the addition completes. This scheme requires more area than a standard adder, but exhibits reduced latency. For hardware simplicity and logic limitation, our FPA hardware does not support earlier LD prediction.

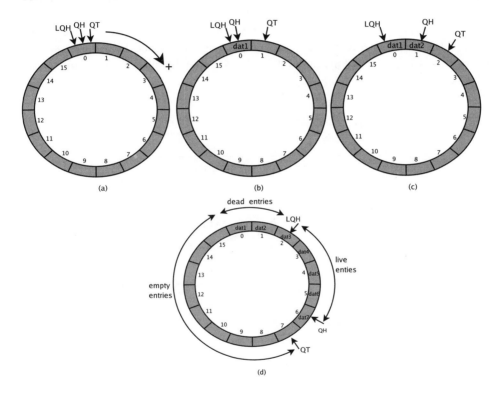

Fig. 6.8   Circular queue-register (QREG) structure. (a) initial QREG state; (b) QREG state after writing the first 32 bit data (dat1); (c) QREG state after writing the second data (dat2) and consuming the first 32 bit data (dat1); (d) QREG state with LQH pointer update and different regions.

6.3.4.2   *Floating Point Multiplier Implementation*

Fig. 6.7 shows the data path of the FMUL unit. As with other conventional architectures, QC-2's FMUL operation is much like integer multiplication. Because floating point numbers are stored in sign-magnitude form, the multiplier needs only to deal with unsigned integer numbers and normalization. Similar to the FALU, the FMUL unit is a three stages pipeline that produces a result on every clock cycle. The bottleneck of this unit was the $24 * 24$ integer multiplications.

The first stage of the floating-point multiplier is the same denormalization module used in addition to insert the implied 1 to the mantissa of the operands. In the second stage, the mantissas are multiplied and the exponents are added. The output of the module is registered. In the third stage, the result is normalized or rounded.

The multiplication hardware implements the radix-8 modified Booth algorithm. Recoding in a higher radix was necessary to speed up the standard Booth multiplications algorithm since greater numbers of bits are inspected and eliminated during each cycle. It effectively reduces the total number of cycles necessary to obtain the product. In addition, the radix-8 version was implemented instead of the radix-4 version because it reduces the multiply array in stage 2.

6.3.5   *Circular Queue-Register Structure*

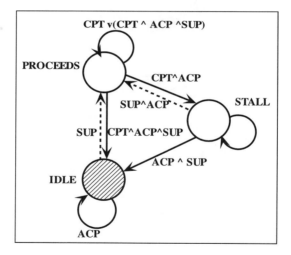

Fig. 6.9   Finite state machine transition for QC-2 pipeline synchronization. The following conditions are evaluated: next stage can accept data (ACP), previous pipeline stage can supply data (SUP), last cycle of computation (CPT).

The QREG structure is shown in Fig. 6.8. For clarity, only the first sixteen entries ($ENTRY_0$ to $ENTRY_{15}$) are shown. Fig. 6.8.(a) shows the QREG initial state. In this state, the QREG is empty and the QH, QT and LQH pointers point to the same location (QREG logical entry 0). When the first data is written into the QREG storage, the QT is incremented by

1. Since no data is consumed yet, the QH and LQH still point to the initial location. This scenario is shown in Fig. 6.8.(b). The QH pointer is incremented by 1 after data 1 (*dat1*) is consumed as shown in Fig. 6.8.(c). Because *dat1* maybe reused again, the LQH pointer still points to $ENTRY_0$, which holds data 1 (*dat1*). A special instruction, named *stplqh* (stop LQH) was implemented to stop the automatic movement of LQH. The automatic LQH movement (default setting) is restored with another special instruction, named *autlqh* (automatic LQH). In some situation the QREG storage may have three types of entries as shown in Fig. 6.8.(d). These entries are: *dead entries* – data is no longer needed, *live entries* – data can be reused, and (3) *empty entries* – no data in these entries.

6.4 Synthesis of the QC-2 Core

6.4.1 *Design Approach*

To make the QC-2 design easy to debug, modify and adapt, we decided to use a high-level description, which was also used by other system designers, such as works in [Micheli (2001); Sheliga (1996); Sharma (1993); Kim (2003); Maejima (1997); Takahashi (1997)]. We have developed the QC-2 core in Verilog HDL. After synthesizing the HDL code, the designed processor has characteristics that enable investigation of the actual hardware performance and functional correctness. It also gives us the possibility to study the effect of coding style and instruction set architectures over various optimizations. For the QC-2 processor to be useful for these purposes, we identified the following requirements:

(1) high-level description: the format of the QC-2 description should be easy to understand and modify;
(2) modular: to add or remove new instructions, only the relevant parts should have to be modified. A monolithic design would make experiments difficult; and
(3) the processor description should be synthesizable to derive actual implementations.

Table 6.1 Normalized code sizes for various benchmark programs over different target architectures

Benchmark	MIPS16	ARM	Thumb	x86	QC-2
H.263	58.00	83.66	80.35	57.20	41.34
MPEG2	53.09	78.40	69.99	53.22	36.75
Susan	47.34	80.48	77.54	46.66	35.12
AES	51.27	86.67	69.59	44.62	34.93
Blowfish	54.59	86.38	82.76	57.45	45.49
FFT	58.09	100.74	92.54	46.27	36.77
Average	**53.73**	**86.05**	**78.79**	**50.9**	**36.77**

The QC-2 has been designed with a distributed controller to facilitate debugging and future adaptation for specific application requirements since we target embedded applications.

This distributed controller approach replaces a monolithic controller which would be difficult to adapt. The distributed controller is responsible for pipeline flow management and consists of communicating state machines found in each pipeline.

In this design, we have decided to break up the unstructured control unit to small, manageable units. Each unit is described in a separate HDL module. That is, instead of a centralized control unit, the control unit is integrated with the pipeline data path. Thus, each pipeline stage is mainly controlled by its own simple control unit. In this scheme, each distributed state machine corresponds to exactly one pipeline stage, and this stage is controlled exclusively by its corresponding state machine. Overall flow control of the QC-2 processor is implemented by cooperation of the control units in each stage based on communicating state machines. Each pipeline stage is connected to its immediate neighbors, and indicates whether it is able to supply or accept new instructions. Communication with adjacent pipeline stages is performed using two asynchronous signals, *AVAILABLE* and *PROCEED*. When a stage has finished processing, it asserts the *AVAILABLE* signal to indicate that data is available to the next pipeline stage. The next stage will, then, indicate whether it can proceed these data by using the *PROCEED* signal.

Since all fields, necessary to find what actions are to be taken next, are available in the pipeline stage (for example operation status ready bits and synchronization signals from adjacently stages), computing the next stage is simple. The state transition of a pipeline stage in the QC-2 is illustrated in Fig. 6.9. This basic state machine is extended to cover the operational requirements of each stage, by dividing the *PROCEED* state into sub states as needed. An example is the implementation of the Queue computation stage, where *PROCEED* is divided into sub states for reading initial addresses values, calculating next addresses values, and addresses fixup (when needed).

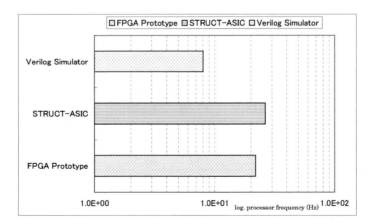

Fig. 6.10   Achievable frequency is the instruction throughput for hardware implementations of the QC-2 processor. Simulation speeds have been converted to a nominal frequency rating to facilitate comparison.

We have synthesized the QC-2 core for Stratix FPGAs and HardCopy devices with Altera Quartus II professional edition [Altera (2008)] tool. In order to estimate the impact of the description style on the target FPGAs efficiency, we have explored logic synthesis for FPGAs. The idea of this experiment was to optimize critical design parts for speed or resource optimizations. In this work, our experiments and the results described are based on the Altera Stratix architecture [Lewis (2002)]. We selected Stratix FPGAs device because it has good tradeoffs between routability and logic capacity. In addition it has an internal embedded memory that eliminates the need for external memory module and offers up to 10 Mbits of embedded memory through the TriMatrix TM memory feature. We also used Altera Quartus II professional edition [Altera (2008)] for simulation, placement and routing. Simulations were also performed with Cadence Verilog-XL tool [Cadence (2008)].

Table 6.2   Execution time and speedup results

Benchmark	PQP-S	QC-2	Speedup
H.263	25980	11777	2.21
MPEG2	22690	10412	2.18
Susan	11321	7613	1.49
AES	5132	1438	3.57
Blowfish	5377	3044	1.77
FFT	9127	5234	1.74

6.5   Results and Discussions

6.5.1   *Execution Speedup and Code Analysis*

Before describing the QC-2 synthesis results, we first present the execution time, speed up and programs size (binaries) evaluation results for several benchmark programs. We obtained these results by using our back-end tool (QC2ESTM) and QueueCore/QC-2 compiler [Canedo (2006, 2007)]. The embedded applications are selected from MediaBench [Lee (1997)] and MiBench [Matthew (2001)] suites. The selected benchmarks include two video compressing applications: H.263, MPEG2; one graph processing algorithm: Susan; two encryption algorithms: AES, Blowfish; and one signal processing: FFT.

Table 6.1 shows the normalized code size of several benchmark programs compiled with a port of GCC 4.0.2 for every target architecture. We selected MIPS I ISA [Kane (1992)] as the baseline and include other three embedded RISC processors and a CISC representative. The last column shows the normalized code size for the applications compiled using the QC-2 compiler [Canedo (2006, 2007)]. The table shows that the binaries for the QC-2 processor are about 70 % smaller than the binaries for MIPS and about 50 % smaller than ARM [Patankar (1999)]. Compared to dual-instruction set embedded RISC proces-

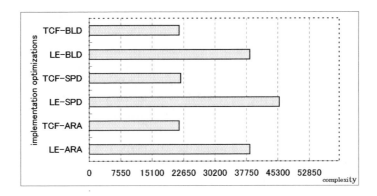

Fig. 6.11   Resource usage and timing for 256*33 bit QREG unit for different coding and optimization strategies.

sors, MIPS16 [Kissel (1997)] and Thumb [Goudge (1996)], QC-2 binaries are about 20 % and 40 % denser, respectively. When compared to the CISC architecture, Pentium processor [Alpert (1993)], QC-2 binaries are about 14 % denser.

Table 6.2 shows the execution time in cycles for serial (PQP-S) and parallel (QC-2) architectures. The last column in the table shows the speedup of the parallel execution scheme over serial configuration. This table shows that the queue computation model extracts natural parallelism found in programs speeding up these embedded applications by factors from 1.49 to 3.57.

6.5.2   Synthesis Results

Table 6.1 shows the hardware configuration parameters of the designed QC-2 core. Table 6.2 summarizes the synthesis results of the QC-2 for the Stratix FPGA and HardCopy targets. The complexity of each module as well as the whole QC-2 core are given as the number of logic elements (LEs) for the Stratix FPGA device and as the total combinational functions (TCF) count for the HardCopy device (Structured ASIC). The design was optimized for balanced optimization guided by a properly implemented constraint table. We also found that the processor consumes about 95.3 % of the total logical elements of the target device.

The achievable throughput of the 32-bit QC-2 core on different execution platforms is shown in Fig. 6.10. For the hardware platforms, we show the processor frequency. For comparison purposes, the Verilog HDL simulator performance has been converted to an artificial frequency rating by dividing the simulator throughput by a cycle count of 1 CPI. This chart shows the benefits which can be derived from direct hardware execution using a prototype when compared to processor simulation. The data used for this simulation are based on event-driven functional Verilog HDL simulation [Ben-Abdallah (2006)].

Table 6.3   QC-2 Hardware Configuration Parameters

Items	Configuration	Description
IW	16-bit	instruction width
FW	8 bytes	fetch width
DW	8 bytes	decode width
SI	85	supported instructions
QREG	256	circular queue-register
ALU	4	arithmetic logical unit
LD/ST	2	load/Store unit
BRAN	1	branch unit
SET	4	set unit
MUL	1	Multiply unit
FPU	2	Floating-point unit
GPR	16	general purpose registers
MEM	2048 word	PROG/DATA memory

Table 6.4   QC-2 processor design results: modules complexity as LE (logic elements) and TCF (total combinational functions) when synthesized for FPGAs (with Stratix device) and Structured ASIC (HardCopy II) families

Descriptions	Modules	LE	TCF
instruction fetch unit	IF	633	414
instruction decode unit	ID	2573	1564
queue compute unit	QCU	1949	1304
barrier queue unit	BQU	9450	4348
issue unit	IS	15476	7065
execution unit	EXE	7868	3241
queue-register unit	QREG	35541	21190
memory access	MEM	4158	3436
control unit	CTR	171	152
QC-2 core	**QC-2**	**77819**	**42714**

The critical path of the QC-2 core with 16 registers configuration is 44.4 ns, that was 22.5 MHz of clock frequency. For QC-2 core with 256 registers, the critical path is 39.2 ns. The clock frequencies for both configurations are low due to the fact that, we synthesized the processor library to random logic of standard cell. However, the performance may be much more improved by using specific layout generation tools.

Fig. 6.11 compares two different target implementations for 256×33 QREG for various optimizations. Depending on the target implementations device, either logic elements (LEs)

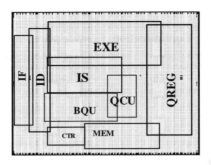

Fig. 6.12   Floorplan of the placed and routed QC-2 core.

or total combinational functions (TCF) are generated as storage elements. Implementations based on HardCopy device, which generates TCF functions give almost similar complexity for the three used optimizations — area (ARA), speed (SPD) and balanced (BLD). For FPGA implementation, the complexity for SPD optimization is about 17 % and 18 % higher than that for ARA and BLD optimizations respectively.

Figure 6.12 shows the floorplan of the placed and routed QC-2 core. The modules of the processor show considerable overlap as logic is mapped according to interconnect requirements.

6.5.3   *Speed and Power Consumption Comparison with Synthesizable CPU Cores*

Queue computing and architecture design approaches take into account performance and power consumption considerations early in the design cycle and maintain a power-centric focus across all levels of design abstraction. In QC-2 processor, all instructions designed are fixed format 16-bit words with minimal decoding effort. As a result, the QC-2 architecture has much smaller programs than either RISC or CISC machines. As we showed in the previous section, programs sizes for our architecture are found to be 50 to 70 % smaller than programs for conventional architectures. The importance of the system memory size translates to an emphasis on code size since data is dictated by application. Larger memories mean more power, and optimization power is often critical in embedded applications. In addition, instructions of QC-2 processor specify operands implicitly. This design decision makes instructions independent from the actual number of physical queue words (QREG). Instructions are, then, free from false dependencies. This feature eliminates the need for register renaming unit, which consumes about 4 % of the overall on-chip power in conventional RISC processors [?Bisshop (1999)].

Performance of QC-2 in terms of speed and power consumption is compared with various synthesizable CPU cores as illustrated in Table 6.3. The SH-2 is a popular Hitachi SuperH based instruction set architecture [IEEE (1997); Hitachi (2007)]. The SH-2 has RISC-type instruction sets and 16x32 bit general purpose registers. All instructions have 16-bits fixed

Table 6.5 Speed and power consumption comparisons for various Synthesizable CPU cores over speed (SPD) and area (ARA) optimizations. This evaluation was performed under the following constraints: (1) Family: Stratix; (2) Device: EP1S25F1020; (3) Speed: C6. The speed is given in MHz.

Cores	Speed (SPD)	Speed (ARA)	Average Power(mW)
PQP	22.5	21.5	120
SH-2	15.3	14.1	187.5
ARM7	25.2	24.5	22
LEON2	27.5	26.7	458
MicroBlaze	26.7	26.7	135
QC-2	25.5	24.2	90

length. The SH-2 is based on 5 stages pipelined architecture, so basic instructions are executed in one clock cycle pitch. Similar to our QC-2 core, the SH-2 also has an internal 32-bit architecture for enhanced data processing ability. LEON2 is a SPARCV8 compliant 32-bit RISC processor. The power consumption values are based on Synopsis software based on reasonable input activities. ARM7 is a simple 32-bit RISC processor and the power consumption values are manufacturer given for hard core. The MicroBlaze core is a 32-bit soft processor. It features a RISC architecture with Harvard-style, separate 32-bit instruction and data buses [Xilinx (2006)].

From the result shown in Table 6.3, the QC-2 processor core shows better speed performance for both area and speed optimizations when compared with SH-2, PQP and ARM7 (hard core) processors. The QC-2 has higher speed for both SPD and ARA optimizations when compared with SH-2 processor (about 40 % for speed optimization and 41.73 % for area optimization). QC-2 core also shows 25 % less power consumption when compared with PQP and consumes less power than LEON2 and MicroBlaze processors. However, QC-2 core consumes more power than ARM7 processor, which also has less area than PQP and QC-2 for both speed and optimization (not shown in the table). This difference comes from the small hardware configuration parameters of ARM7 when compared to our QC-2 core parameters.

6.6 Conclusions

In this chpater we presented the architecture, design and evaluation of a produced order queue processor with single floating point support (QC-2) and a novel technique used to extend immediate values and memory instruction offsets. The QC-2 Core is targted for embedded single and multicore architecture. Evaluation results reveal that the QC-2 processor achieves a speed of about 25.5 and 22.5 MHz for QREG16 (QREG size is 33*16 entries) and QREG256 (QREG size is 33*256 entries) respectively. We also found that the processor consumes about 95.3 % of the total logic elements (LEs) of the Stratix EP1S80B9 device. As a result, it fits on a single Stratix device with an internal embedded memory that eliminates the need for external memory module, thereby obviating the need to perform multi-chip partitioning which results in a loss of resource efficiency. Only a few clearly identified components, such as the Barrier and the QREG units, need to be specially optimized at the HDL source level to achieve efficient resources usage. From the comparison results, we also conclude that the QC-2 processor core shows better speed performance for both area and speed optimizations when compared with SH-2, PQP and ARM7 (hard core) processors. On average the QC-2 has about 40.87 % higher speed than SH-2 processor. QC-2 core also shows 25 % less power consumption when compared with PQP, and consumes less power than SH-2, LEON2 and MicroBlaze cores.

Chapter 7

Reconfigurable Multicore Architectures[1]

With the proliferation of portable devices, new multimedia-centric applications are continuously emerging on the consumer market. These applications are pushing computer architecture to its limit considering their demanding workloads. In addition, these workloads tend to vary significantly at run time as they are driven by a number of factors such as network conditions, application content, and user interactivity. Most current hardware and software approaches are unable to deliver executable codes and architectures to meet these requirements. There is a strong need for performance-driven adaptive techniques to accommodate these highly dynamic workloads. This chapter shows the potential of these techniques in both software and hardware domains by reviewing early attempts in dynamic binary translation on the software side and FPGA-based reconfigurable architectures on the hardware side. It puts forward a preliminary vision for unifying runtime adaptive techniques in hardware and software to tackle the demands of these new applications. This vision will not be possible to realize unless the notorious reconfiguration bottleneck familiar in FPGAs is addressed. The chapter concludes by pointing several future directions to explore in order to realize the full potential of runtime adaptation.

7.1 Introduction

In recent years, the proliferation of portable devices has changed user demands in the electronic consumer market where new applications such as multimedia applications have emerged as a driving force of innovation in this market. These devices are evolving into portable systems with a rich functionality to support these applications. They can run small operating systems and support concurrency to execute several applications at the same time. Users can use these devices to view a video clip, set an appointment in their personal calendar, and answer an incoming phone call at the same time. Most multimedia-centric applications such as video entertainment and games display requirements and characteristics that are quite different from those found in the applications which appeared before the advent of the multimedia age. These characteristics include [Diefendorff (1997)]:

(1) Real-time responses: Most processing tasks such as video encoding/decoding in multimedia applications must meet real time constraints in order to give a realistic appearance from the user's perspective.

---

[1] Abdel Ejnioui, University of South Florida Polytechnic

(2) Continuous data types: Multimedia applications tend to operate on continuous data types that are quite different from the data types manipulated before the advent of multimedia centric applications. As such, these data types have expanded from 32 to 64 bits in order to accommodate the large numerical ranges of these applications. However, the values of these data tend to vary widely at run time.

(3) Fine-grain data parallelism: These applications exhibit data streams which can be broken into large collections of small data elements such as pixels or voxels before these elements can undergo identical processing steps.

(4) Coarse-grain parallelism: These applications tend to execute on a number of concurrent threads. For example, video entertainment requires separate threads to perform video encoding/decoding and audio encoding/decoding,

(5) Dominance of iterative kernels: At the heart of multimedia applications are signal and image processing algorithms. These algorithms are dominated by the occurrence of a number of compute-intensive loops with locally referenced data in space and time.

Furthermore, these characteristics tend to change while the application is running based on a number of environmental factors. For instance, the quality of a video stream can vary significantly based on network bandwidth. While this bandwidth can be mostly stable in wired networks, it is not so in wireless networks. In addition, this quality can also be affected by device power and resources constraints if the stream is delivered to a mobile phone or PDA [Liu (2005)]. Recently, popular 3D graphic games that were played on networked desktop computers are migrating to mobile devices. The 3D contents of these games are distributed over the network to desktop computers and mobile devices alike [Micheli (2001)]. Mobile devices will be forced to reduce the level of detail when drawing 3D objects in order to provide acceptable frame rates in these games [Tack (2004)]. In addition, 3D games workload can exhibit in general over an order of magnitude of interframe variations as scenes with considerable details are frequently followed by relatively empty scenes [Lafruit (2000)]. These runtime workload variations can significantly affect performance. As a result, applications must have the capability to select appropriate performance parameters in order to sustain a reasonable quality from the user's perspective.

7.1.1 *Performance-Driven Approaches in Software*

Previous approaches in the software or hardware domain to support performance-demanding applications will not be able to meet the demands of these new applications. For instance, compile time approaches for parallelism are completely inadequate for these new applications. These approaches have been focusing either on compilers or operating systems. While existing compilers are efficient at extracting instruction-level parallelism, they have yet to show any significant progress in extracting task-level parallelism [Asanovic (2006)]. Recent innovations in exposing parallelism such as access optimization to cache and removing cache conflicts for maximum cache sharing are considered marginal at best [Fisher (1997)]. Assuming that major breakthroughs can be realized in synthesizing optimized code for task-level parallelism, the engineering effort needed to integrate the implementation of these breakthroughs may be difficult to justify. In fact, the number of lines of code in these compilers has increased to the point where it is becoming difficult to re-engineer their internal data structures and algorithms. On the operating systems side, the picture is not promising either. While operating systems in the server world are monolithic, they remain to some extent minimal in the embedded world. However, the server world had to integrate virtualization with operating systems to sustain application performance while overcoming the difficulties associated with farming a large diversity of hardware and software resources. Embedded systems have yet to take advantage of virtualization when their hardware resources are becoming noticeably diverse to meet the requirements of multimedia applications in addition to the increasing appetite for consuming these resources by these applications. Nowadays, it is not uncommon to encounter multi-gigabyte file systems and complex web browsers in portable devices [Asanovic (2006)]. What is needed is a form of runtime adaptation in which these applications can alter their runtime environment to meet user's demands or satisfy resource constraints. Among previously proposed approaches, virtualization seems to be the best approach to support runtime adaptation. Virtualization can be accomplished in the software or hardware domain. In the software domain, many believe that there should be more focus on runtime (i.e., dynamic) techniques of compilation and optimization.

7.1.2   *Performance-Driven Approaches in Hardware*

To address runtime adaptation from a system-level perspective, it is imperative to consider its potential in the hardware domain also. As a matter of fact, runtime adaptation can be supported in the hardware domain through architectural innovation thanks to the increasing scale of integration made possible by nanometer CMOS technology. The demanding performances of multimedia-centric applications cannot be supported without powerful processors. However, computer architects recently came to the realization that uni-processor systems cannot deliver the performance necessary to support these applications. Advances in clocking strategies and increasing reliance on co-processors are becoming inadequate to meet the computing demands of these applications. As a result, many advocate the exploration of multi-processor architectures to meet these demands. Since current CMOS technology provide unparalleled scales of integration, it is possible to place these multiprocessor systems on a single die leading to an entire system-on-chip (SOC). In fact, many vendors have already introduced in the market multithreading multiprocessor systems for multimedia applications. For instance, nVIDIA offers currently a 128-multiprocessor computing engine inside the Tesla device for simulation and graphic-rich applications [nVIDIA (2007)]. While multiprocessor systems can be an excellent medium for mapping applications rich in task-level parallelism, they cannot sometimes prevent performance degradation in case compute-intensive tasks do not match the available processors [Gohringer (2008,?)]. This inevitable limitation is due to the fact that multiprocessor SOCs have a fixed architecture which cannot adapt at runtime to the demands of the application. A possible approach to overcome this limitation is to rely on runtime reconfiguration. Hence, the attractiveness of reconfigurable architectures, namely Field-Programmable Gate Arrays (FPGAs) [Coffer (2006)] and Coarse-Grain Configurable Arrays (CGRAs) [Hartenstein (2001)].

7.1.3 *Potential of FPGA SOCs*

CGRAs such as KressArray [Hartenstein (2000)], PipeRench [Sheliga (1996)], Garp [Callahan (2000)] are reconfigurable arrays with regular wide datapaths. The objective of using wide datapaths in CGRAs is to simplify routing and minimize reconfiguration overhead. Although CGRAs were successfully used to accelerate a number of applications, they nevertheless lose significant flexibility by fixing more or less the structures of their datapaths [Marchall (1999); Alsolaim (2000)]. In contrast, FPGAs contain fine-grain reconfigurable fabric which can be reconfigured to support any datapath regardless of its width although at the expense of higher reconfigurable overhead. This high degree of flexibility or versatility provided by FPGAs make them perfect candidates to support runtime adaptation in hardware. Any amount of structure fixing would be an impediment to runtime adaptation if the fixed structure does not match the computation to be mapped on it. Even if coarse-grain datapaths are needed to support a given computation, newer large capacity FPGAs can support these computations with resources that can be found in CGRAs. Many modern FPGAs, such as Virtex chips, can provide a gate capacity in the millions supported by a rich mix of coarse- and fine-grain reconfigurable resources. In fact, these FPGAs can contain hardwired processors (e.g., PowerPC), multipliers, memory blocks, and LUT-based reconfigurable logic. From the perspective of hardware runtime adaptation, FPGAs are the perfect technology to combine with software runtime adaptation.

To show the potential of runtime adaptation, Section 7.2 presents three different architectures which are good examples of runtime hardware adaptation. Section 7.3 presents two different forms of software runtime adaptation. Section 7.4 presents possible new directions in runtime adaptation to address the computational requirements of newer applications. Section 7.4 concludes the chapter.

7.2   Runtime Hardware Adaptation

Currently, FPGAs are the technology of choice for supporting run time adaptation of high performance architectures. Many FPGA vendors such as Xilinx offer large capacity chips that contain a variety of reconfigurable resources [Xilinx (2009); Altera (2009)]. These resources can be reconfigured wholly or partially without interrupting the operations of other resources inside the chip. Considering their capacities and capabilities, these FPGAs are full blown reconfigurable SOCs whose resources can be exploited for run time adaptability. Using this technology, a variety of processor-based architectures can be mapped at various times to sustain the performance of data intensive applications. To increase the degree of parallelism, a hardware architecture can be modified wholly or partially to adapt to the computing demands of the application. This adaptation can be achieved in three different ways at the hardware level: processor array architectures, datapath array architectures, and single processor architectures.

7.2.1 *Processor Array Architectures*

Processor array architectural adaptation relies on the availability of an already designed and optimized architecture for a given application. This architecture can be mapped on the reconfigurable SOC at run time by specifying a set of parameters such as the number of processors and the communication structure used between the processors. The architecture can be altered to sustain the computing demands of the application by changing the bitwidth of the processors or extending their instruction sets. Alternatively, the communication structure of the architecture can be altered instead of changing its processors. A popular architecture for this adaptation is a multiprocessor architecture consisting of a number of general purpose or Application Specific Instruction Set Processors (ASIPs) [Jacome (2000)] connected by a network-on-chip (NOC) [Bjerregaard (2006)]. General purpose multiprocessor architectures have recently begun to appear in many desktops and servers [Intel (2006); Wikipedia (2009); Barreh (2006)].

The RAMPSoC approach is a typical example of this adaptation [Gohringer (2008,?)]. The objective in RAMPSoC is to accelerate through parallelism data intensive applications by using dynamic reconfiguration to adapt multiprocessor architecture and communication bandwidth. This is achieved by exploiting concurrency at data, task, and instruction levels within the application. From an architectural perspective, RAMPSoC looks for ways to exploit runtime adaptation at the SOC, communication, and processor level using the concurrency patterns discovered at data, task, and instruction level. Figure 7.1 shows the three levels of hierarchy of architectural adaptation.

At the SOC level, consideration of performance and power requirements drives an application-level process of coarse task partitioning in order to assign partitioned tasks to specific processors. At the communication level, a structure such an NOC, bus, or point-to-point is synthesized based on the task interconnection patterns within the application. At the processor level, individual processor architectures are customized while their instructions are optimized to adapt to processor customization. Mostly, these processors are Reconfigurable Instruction Processors (RISPs) [Barat (2002)] since they can be highly adaptable and more power efficient than ASIPs. In addition, their adaptability can be increased by swapping at run time the contents of their instruction and data memory. A target application that can benefit from template architectural adaptation is real-time image processing which consists of object recognition and tracking tasks from input video streams at 25 to 50 Hz frame rates. For this application, a heterogeneous multiprocessor architecture is synthesized consisting of three types of processors and a finite state machine connected by a switch-based NOC as shown in Figure 7.2.

While the three types of processors are different in terms of bitwidth and performance, they can be augmented by one or more accelerators used for extending their instruction sets. If

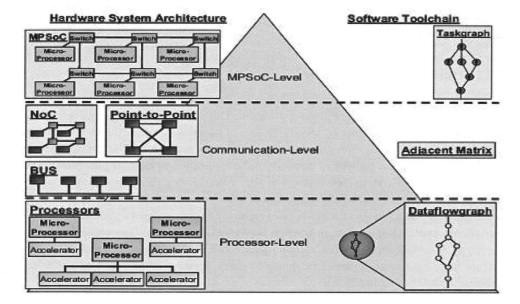

Fig. 7.1   The three architectural levels in RAMPSoC

the processing requirements of the running application changes dynamically, the accelerators can be swapped in and out as in the case when an image processing application needs
a different filter based on lighting conditions. In addition, if the application reads data from
different sensors, it may be necessary to alter the micro-architecture of the processor and
the NOC when data with different widths is fed to the SOC. At specific time intervals, the
application may need to run in low power mode. In this case, run time reconfiguration can
be used to shut down a processor or remove it altogether from the reconfigurable SOC.
While these multiprocessor architectures can be quite efficient in supporting parallel execution, they may not perform as well if careful consideration is not given to the NOC.
Figure 7.3 shows a fictive mesh-based NOC.

In fact, the bitwidths of network connections can affect significantly the throughput of the
architecture. As a result, run time reconfiguration of critical components in the NOC is necessary to allow the NOC to sustain the needed data throughput within the reconfigurable
SOC. For instance, network switches must be highly flexible to support data transfers to
and from the diversity of processors attached to these switches. To avoid routing congestion, a multi-layer NOC is highly desirable since it can increase the number of possible
paths between network nodes. In addition, the NOC must be highly malleable to allow its
topology to change as processors are added or removed from the architecture. A possible
realization of this NOC is shown in Figure 7.4 where the role of the ModCom unit is to
connect a network switch to a module. The latter can be any functional resource such as a
processor, an accelerator, or an FSM.

While large capacity reconfigurable SOCs contain hardware components to support recon-

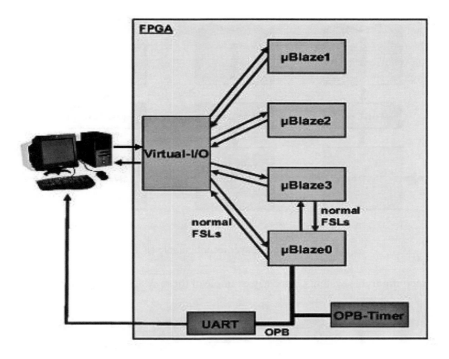

Fig. 7.2   An example of a heterogeneous multiprocessor architecture template for a real-time image processing application

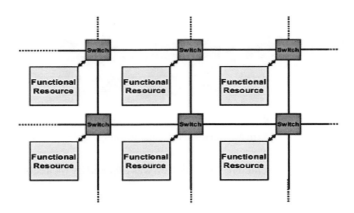

Fig. 7.3   Mesh-based NOC

figuration, the process of reconfiguring these chips is not trivial. Most vendors of these chips do not offer any advanced tools to support reconfiguration at run time. Early attempts at supporting run time reconfiguration through software proposed space-inefficient

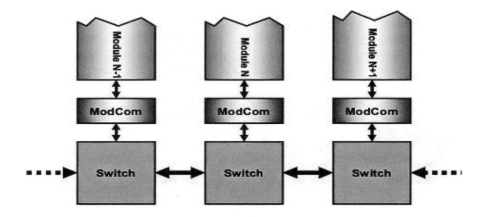

Fig. 7.4   A multilayer reconfigurable NOC

dynamic reconfigurable systems. An early example of these systems is shown in Figure 7.5.

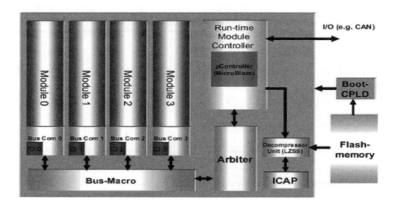

Fig. 7.5   Vertically oriented dynamic reconfigurable systems

Preference for vertically-oriented reconfiguration was dictated by the reconfiguration architecture of reconfigurable SOCs such as the Virtex FPGAs [Xilinx (2009)]. This architecture made re-sizing and orienting dynamically reconfigurable areas cumbersome and space
wasteful [Becker (2006)]. As shown in Figure 7.5, the vertical slots fixed for dynamic reconfiguration have a fixed size, which can be problematic if the computational task to be
loaded is too large to fit in the slot or too small to occupy most of the slot area. It be-
came clear that an infrastructure with finer granularity control over the sizes and locations
of dynamically reconfigurable areas is necessary to support run time adaptation. Such an
infrastructure must take advantage of the 2 dimensional nature of the reconfigurable fabric

of the reconfigurable SOC. Figure 7.6 shows an example of a dynamically reconfigurable system based on such an infrastructure.

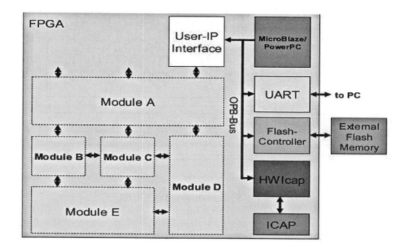

Fig. 7.6   A 2-dimensional dynamic reconfigurable system

This approach has the advantage of loading computing tasks of different sizes to accommodate time-varying application workloads.

7.2.2   Datapath Array Architectures

In datapath array architectural adaptation, an array of functional units or processing elements (PEs) such as adders and multipliers is available to boost the performance of the application. This architecture is intended to be mapped on available large capacity FPGAs. The array is loaded at runtime to accelerate the execution of the application. The operations of each PE and the communication link between PEs can be reconfigured every clock cycle. However, the entire array can be reconfigured into a different array to support a different application. The best example of this approach is the QUKU architecture [Shukla (2006,?,?, 2007)]. Figure 7.7 shows a diagram of QUKU.

This array can be mapped at compile time by mapping computational tasks onto PEs. These tasks are obtained from a dataflow graph representing the application flow of data and operations as the design flow of QUKU shows in Figure 7.8.

Figure 7.9 shows an example mapping of an FIR filter. Needless to say that this architectural adaptation is ideal for data intense applications such as filters and transforms. Because of the nature of its mapping, this architecture provides two levels of application-specific reconfigurability. At the lowest level, reconfiguration can take place across the entire SOC to upload a new array in order to accelerate an application which is about to begin.

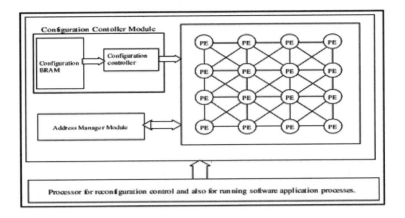

Fig. 7.7   System diagram of QUKU

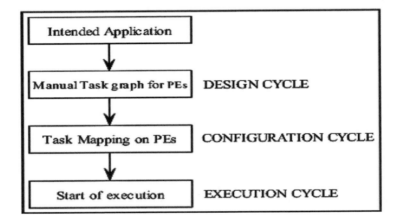

Fig. 7.8   Design flow of QUKU

At the highest level, reconfiguration can take place at a PE or an inter-PE link for maximum reuse of arithmetic and logical operators. While reconfiguration at the lowest level can be in the order of several milliseconds, reconfiguration at the highest level can take only a few nanoseconds. Configuration codes can be generated for each PE in the array form the task graph obtained in the design cycle of QUKU. To configure PEs in the array, they are addressed from left to right and top to bottom. The configuration code consists of 16-bit words and 10-bit recurrence counter values. An example of configuration codes is shown Table 1 below for the task graph shown in Figure 7.9. This architecture can achieve relatively fast reconfiguration times at the highest level as shown in the table below. Table 7.2 compares the bit size of the configurations and their speed for QUKU and the Montium

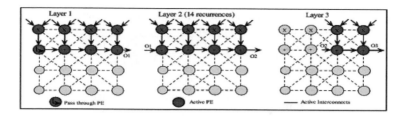

Fig. 7.9   Temporal and spatial mapping of an FIR filter on QUKU

processor [Heysters (2004)]. The latter is a coarse-grain general purpose reconfigurable array. In terms of performance, QUKU's performance is located between the performance of a MicroBlaze soft processor and a custom core with eight time speed improvement over the performance of the MicroBlaze processor. In addition, its area cost is much lower for the performance it delivers compared to the area cost of the custom core. Figure 7.10 shows the area cost and clock performance for the implementation of an FIR on QUKU, MicroBlaze, and custom core.

Table 7.1   Configuration codes for the task graph of Figure 10

Config word	Explanation
800000FF	PE 0-7 reset indication
03480001	Multiplier configuration
8000000F	Valid for PE 0-3
02540001	PE 4 configured as a pass throuth element
80000010	
04540001	PE 5-7 configured to add the O/P of
800000E0	previous adder and O/P of their multiplier
0414000E	PE 4-7 configured to add the O/P of
800000F0	previous adder and their result
04940001	PE 6 configured to add the O/P of
00000040	previous stage. O2 and O/P of mult 2
04140001	PE 7 configured to add the O/P of
00000080	PE 6 and O/P of mult 3 and indicate final result

This architecture supports parallelism by acting as a virtual hardware platform that can be used to implement applications that are larger than the available hardware. This architectural adaptation has the advantage of relieving the designer from identifying the optimal mix of operators needed to accelerate the application. Instead, its reconfiguration capabilities at the highest level allow it to adapt to various datapath operations needed in the application.

Table 7.2   Comparison of configuration size and
speed in QUKU and Montium

	FIR 5	FIR 20
OUKU	120 bits	146 bits
	10 cycles	12 cycles
Montium	1 968 bits	4 320 bits
	123 cycles	270 cycles

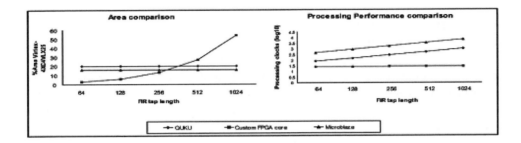

Fig. 7.10   Area and clock performance of QUKU, MicroBlaze, and a custom core for an FIR implementation

7.2.3   *Single Processor Architectures*

In single processor architectural adaptation, a generic soft processor is used to sustain the
application throughput when space is not available to upload computational tasks. This approach takes advantage of the fact that many computational tasks tend to occupy relatively
large areas of the reconfigurable fabric of the SOC while these soft processors do not. A
typical example of this approach is proposed in [Montone (2008)]. In this approach, the
assumed target architecture consists of a static area in charge of controlling the communication structure between pre-set reconfigurable areas as shown in Figure 7.11.

The reconfigurable areas can be reconfigured to implement a state machine, a pipelined
unit, a processing element or any dedicated core. For better integration, the processing element consists of a MicroBlaze soft core which communicates with fixed areas of user logic
through a bus as shown in Figure 7.12. This communication is supported by an on-line pe-
ripheral bus (OPB) bus for compatibility with the MicroBlaze core. This core is equipped
with instruction and data memory. Hence, the name Harvard Architecture of Reconfigurable Processing Elements (HARPE). The user logic areas can be used as accelerators
while the core can act as a controller thus combining software flexibility with hardware
speed. This approach has the distinctive characteristic of reconfiguring the contents of both
memories to take maximum advantage of software flexibility. This approach requires that
the target architecture provides facilities to swap in and out of the SOC cores and user logic
without interrupting the SOC operation.

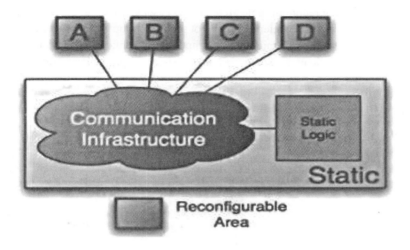

Fig. 7.11   Logical view of the target architecture

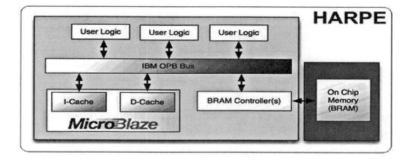

Fig. 7.12   HARPE processing element

A typical scenario in which this approach can sustain application performance is shown
in Figure ??. In this scenario, S denotes the occupied static area in the target architecture
while the white striped area represents the available reconfigurable areas. C and O represent
two incoming dedicated acceleration cores for convolution and open respectively that need
to be uploaded on the SOC. Open implements the combination of erosion and dilatation operations in image processing. It is obvious in 1) on the figure that only one of the incoming
cores can be uploaded as there is not enough reconfigurable area to upload both. One can
opt to upload first C before O as shown in 2) on the figure. If that is the case, two alternatives are possible. The system can wait for C to complete before uploading O as shown in
the left illustration of 3). On the other hand, the system can upload at the same time C and
a HARPE processing element denoted by H as shown on the right illustration in 3) since H

can fit in the remaining reconfigurable area. In this case, H will execute the instruction of O. While a HARPE processing element may not produce an optimum performance similar to that of O, it would at least sustain the throughput of the application without allowing the performance of the application to degrade below a specified performance threshold.

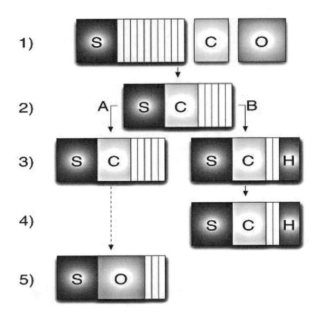

Fig. 7.13   Reconfiguration scenario

This sustained acceleration would not be possible without the realization that the HARPE core is relatively small to fit in smaller reconfigurable areas than the areas occupied by other dedicated computing cores as illustrated in Table 7.3 [Kulkarani (2002)].

While in specialized architectural adaptation, a template architecture is modulated to exploit parallelism during run time, this adaptation relies on a single area-efficient processing element to exploit small reconfigurable areas to sustain system performance. In terms of flexibility, demand-driven adaptation can be adopted to support a variety of applications as long as a library of dedicated cores is available.

7.3   Summary of Hardware Adaptation

Table 7.4 presents a summary of the architectural adaptations presented above. The architectures are contrasted based on specific characteristics that are critical to run time adaptation.

Table 7.3   Area cost of soft-core processors and dedicated cores

Core	Area usage (LUTs)	Area usage (%)
MicroBlaze (Area optimized configuration)	1566	15.9
MicroBlaze (Performance optimized configuration)	1679	17
Sobel mag	1943	19.7
Convolution	1783	18.1
Convolution 5	3235	32.8
Convolution sm	2609	26.5
Prewitt mag	1815	18.4
Open	4500	45.7
Close	4500	45.7
Wavelet	2172	22
Trideagonal	7476	75.6

Table 7.4   Summary of hardware architectural adaptation

Characteristics	Hardware Architectural Adaptations		
	Processor Array	Datapath Array	Single Processor
Number of computing components	Multi-processors	Multi PEs	Single processor
Communication structure	NOC	Simple links	No links
Components complexity	RISP core	Funcyional unit	MicroBlaze core
Loadable instances	Single	Single	Several
Target applications	Special	Special	General
Loadability	Pre-runtime	Pre-runtime	At runtime
Customizability	During runtime	During runtime	Before runtime
Support tools	High level	High level	Low level

Column 1 of the table presents these characteristics while the remaining columns specify these characteristics in the three architectures. In the row labeled Communication structure, data transfer and coordination in multiprocessor architectures require precise networking protocols which can be implemented only on NOCs. The row labeled Loadable instances indicates if one or more instances of the architecture can run simultaneously on the SOC. If more than one instance of the architecture can be uploaded, the architecture is not tightly coupled with the rest of the system on the SOC. On the other hand, if several architecture instances can operate at the same time on the reconfigurable fabric of the SOC, a higher degree of concurrency can be achieved at the expense of complex coordination of data transfer and signaling between the application and the architecture instances. Loadability indicates if the architecture can be uploaded on the reconfigurable fabric before or during runtime. An architecture that can be uploaded during runtime has the advantage of providing the running application with a higher degree of runtime adaptation in terms of sustaining performance. Although architectures that can be uploaded before runtime can be customized to some degree during run time, their superiority stems from the fact that they are optimized to boost the performance of the application once they are uploaded on the SOC. Customizability indicates whether the architecture can be modified on the SOC while it is operating. While the single processor architecture is fixed, it is used as a way to prevent performance degradation. In this case, customization will not improve performance. However, processor and datapath arrays are intended to boost performance as workload conditions change. As a result, they must accommodate some limited modification to the granularity of their processors and communication structure. To implement these array architectures, sophisticated compilers are need. These compilers can identify compute-intense kernels within the application that are good candidates for acceleration on the array. These compilers can also map these kernels onto these arrays and synthesize runtime schedules that are efficient in terms of throughput and power consumption.

7.4   Runtime Software Adaptation

Many of multimedia-centric applications are intended for portable devices, personal entertainment, and automotive electronics. These applications are notorious for their stringent requirements in terms of power, performance, reliability, and security. In addition, these applications tend to exhibit unexpected workload variations due to a highly dynamic user and environment interaction as explained in the introduction. However, these applications present numerous opportunities to exploit parallelism at instruction-, data-, and thread-levels. While parallel software is the best way to exploit these forms of parallelism, it also known that such software is difficult to develop and test. Software developers are faced with numerous challenges related to hardware variability and application parallelism. It is no more beneficial to write code and optimize it to run on a specific platform given hardware diversity. What is needed is software that is able to run on hardware in an independent manner. To take advantage of the application's inherent parallelism, this software or part of it must be able to migrate from one hardware to another at different points in time while the application is running in order to meet the demands of its dynamic workload. This new capability can equip the software with a high degree of runtime adaptability that is highly needed to boost the performance of parallel applications.

In the context of reconfigurable architectures, the ability of software to adapt is critical in taking advantage of the potential of these architectures. In fact, the potential of hardware architectural adaptation presented in this chapter would not be tapped without significant software adaptation. To take advantage of processor or datapath array architectures, the system must re-direct at some time the instruction stream from one side of the SOC to these array architectures while translating from one instruction set to another appropriate to these arrays during runtime. This can be possible only by virtualizing code execution across the entire SOC. It would be ideal if this virtualization is completely transparent to the programmer and imposes a minimal overhead on the runtime environment. While virtualization has been explored in several research areas such as programming languages, operating systems, computer clusters, and grid computing, it has yet to show its potential in reconfigurable computing. Among the most promising techniques in virtualization towards runtime software adaptation for reconfigurable architectures is dynamic binary translation (DBT). DBT has been gaining attention recently as a powerful technique for runtime software adaptation [Altman (2009)]. In its broadest meaning, DBT represents any runtime transformation of executable code. This transformation can consist of translating native code from an instruction set architecture to another one, optimizing and instrumenting executable code, as well as compiling just-in-time bytecode into native code [Altman (2001)]. DBT has been applied in reconfigurable systems in two forms: Warp Processing and Dynamic Instruction Merging.

7.4.1 *Warp Processing*

In warp processing (WP), binary executable code is partitioned into a software part to run on a processor and a hardware part destined to run on reconfigurable hardware as shown in Figure 7.14 [Stitt (2002)]. In contrast, most code partitioning approaches operate at the source level. Figure 7.14 shows the partitioning process. While most traditional approaches rely on partitioning before or during compilation of source code as shown in Figure 7.14.(a), binary-level partitioning is performed on binary executable code as shown in Figure 7.14.(b). The application's code is compiled first before it is partitioned. This partitioning approach targets critical regions in the execution of an application [Lysecky (2006)]. These regions tend to be loops or subroutines that are frequently executed to account for 10 % or more of the execution time of an application. It was observed in many applications in embedded systems that a large portion of execution time is caused by a few critical regions [Stitt (2002)]. By accelerating these regions, a significant improvement in performance can be obtained. Once these critical regions are identified, partitioning requires specific tools to perform decompilation, behavioral synthesis, logic synthesis, placement and routing on a reconfigurable architecture.

Since this partitioning process must be done at run time in a transparent and non-intrusive fashion, feasibility studies showed that it is efficient if these tools are implemented on a lean processor. The rationale for this design choice is the realization that these dynamic tools target only critical regions most of which consist only of a few dozen of lines of code resulting in hardware consisting of only 10,000 to 30,000 gates [Lysecky (2004)]. The integration of on-chip tools with a processor and reconfigurable fabric lead to a new

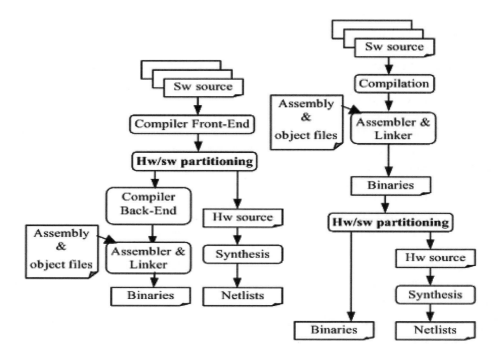

Fig. 7.14   Hardware/software partitioning approaches

hardware architecture called warp processors. Figure 7.15 shows an overview of a warp processor.

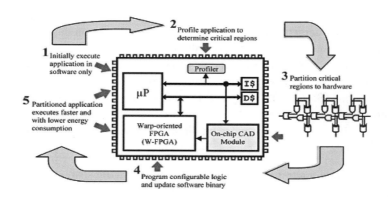

Fig. 7.15   Overview of a warp processor

A warp processor contains a processor, a profiler, on-chip compiling tools, and an opti-

mized reconfigurable fabric called Warp FPGA (W-FPGA). While the application is executing on the processor, the profiler monitors the executable code in order to identify critical
regions in the binary code. This non-intrusive profiler watches instruction addresses on the
instruction memory to detect backward branches which are associated with loops. If such
a branch is detected, the profiler updates an eight-bit entry in a special cache for storing
branch frequencies. The transparency of the profiler is achieved by implementing it as a
2000-gate circuit with a small cache. After profiling, the on-chip CAD module, called
Riverside on-chip CAD (ROCCAD), performs partitioning, synthesis, mapping and routing of the identified critical regions in executable code. Figure 7.16 shows an overview of
ROCCAD tools.

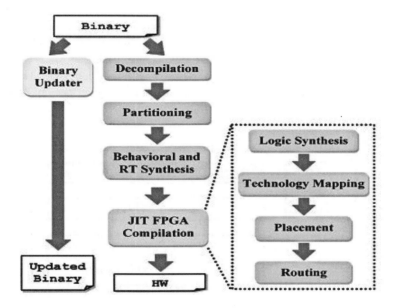

Fig. 7.16   Overview of the ROCCAD tool chain

The first step performed by the ROCCAD tools is decompilation in which each assembly
instruction is converted into an equivalent register transfer (RTL) statement. The latter offers a hardware-independent instruction set. The resulting RTL code is parsed to build a
control graph associated with a set of small data flow graphs which are merged into a single control/data flow graph of the critical region. Next, ROCCAD tools analyze the critical
regions to determine which ones are candidate for hardware implementation by applying
a simple partitioning heuristic. For each critical region that is candidate for hardware implementation, ROCCAD tools synthesize its corresponding control/data flow graph into a
circuit description. The resulting circuit description, in the form of a netlist, is fed to a
just-in-time (JIT) FPGA compiler to map, place, and route the netlist on the W-FPGA of

the warp processor. After this process is complete, the original binary executable needs to be updated accordingly. The binary updater then removes the critical regions implemented in hardware from the binary executable and replaces them with jump instructions to hardware activation codes. The activation code consists of a write to a port connected to the W-FPGA enable signal followed by a small code responsible for powering down the processor into sleep mode. The processor resumes its normal execution only when it hears an interrupt from the W-FPGA. An alternative implementation of the ROCCAD tools is to run them as a minimal software task on the main processor thus sharing resources with the running application. Experiments showed that the place-and-route step in the JIT compilation process is the most computationally expensive step on commercial FPGAs considering the complexity and variety of routing resources available in these FPGAs [Lysecky (2004)]. A router-friendly configurable architecture would be helpful in supporting on-the-fly routing of small circuits. In the same time, this architecture ought to be optimized for supporting the types of operations found in most critical regions of binary code in embedded applications. Figure 18 shows this warp-oriented architecture named above as W-FPGA. A W-FPGA consists of a data-address generator (DADG) with loop control hardware (LCH). In addition, it contains three input-output registers, a 32-bit multiplier-accumulator (MAC), and a routing-oriented configurable logic fabric.

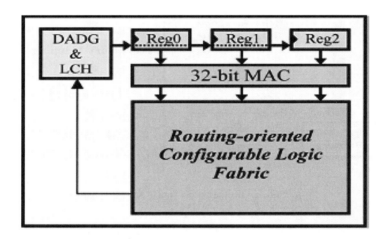

Fig. 7.17   Overview of W-FPGA

The registers can feed input to the MAC or the fabric. In addition, they can operate as output registers to the results produced by the configurable fabric. Because the W-FPGA needs to access memory and control loop execution, the DADG is designed to handle all memory accesses and loop control. The MAC unit is necessary due to the prevalence of multiplications associated with multiply-and-accumulate operations in most critical regions of embedded applications. The configurable fabric consists of configurable logic blocks

(CLBs) surrounded by switch matrices as shown in Figure 19. Each CLB consists of two 3-input 2-ouput lookup tables (LUTs). Such CLBs make mapping and placement less time consuming. On the other hand, a switch matrix routes eight signals on each side where four signals are short wires between adjacent matrices while the other four signals span to each other switch matrix. For experimentation purposes, the warp processor is compared to traditional hardware/software partitioning approaches targeting FPGAs by using 15 benchmarks for embedded systems.

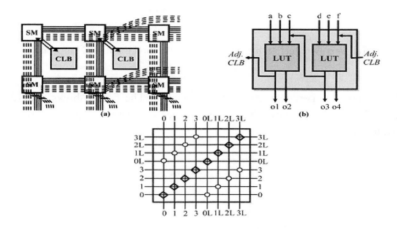

Fig. 7.18   Overview of the routing-oriented configurable logic fabric

Figure 7.19 and 7.20 show the speedup and energy reduction achieved by the warp processor and traditional hardware/software approaches. Figure 7.19 shows that the warp processor can speedup applications on average 6.3 times more than the traditional hardware/software partitioning approach. In addition, the warp processor can reduce energy during execution on average by 56% more than the traditional hardware/software partitioning approach. The area cost of W-FPGA is also evaluated by estimating the amount logic and routing resources within the routing-oriented configurable logic fabric required to implement the critical regions of the benchmarks.

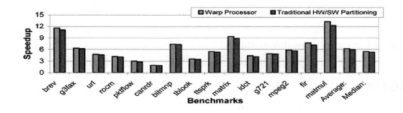

Fig. 7.19   Overall speedup by the warp processor and traditional hardware/software partitioning

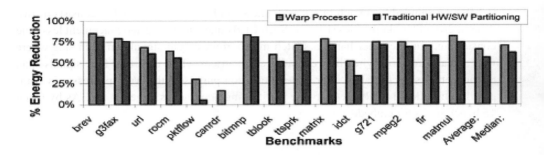

Fig. 7.20   Overall energy reduction by the warp processor and traditional hardware/software partitioning

Figure 7.21 shows the percentage of these resources. The figure shows that 12 % of the available CLBs and 39 of routing resources are needed to implement the critical regions dispatched to the W-FPGA.

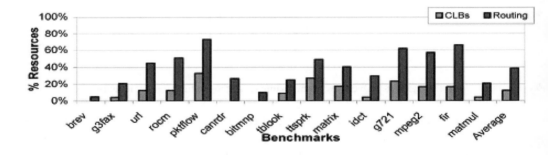

Fig. 7.21   Percentage of resources used to implement the benchmarks in the routing-oriented configurable logic fabric

These results show that warp processing is a promising form of dynamic binary translation from the perspective of runtime adaptation. It has significant potential in serving the performance of embedded parallel applications. However, warp processing appears limited in its applicability of DBT since it focuses mainly on loop-based critical regions consisting of a small number of operations. Recent applications in portable devices tend to have continuous running loop with complex arithmetic operations involving floating point numbers.

7.4.2   Dynamic Instruction Merging

A form of DBT that has been applied to runtime adaptation of executable code in reconfigurable systems is dynamic instruction merging (DIM) [Beck (2008, 2005)]. DIM is suitable for detecting and translating instructions to run on reconfigurable hardware. Because DIM is intended to be transparent to the user, the impact of its execution overhead must be minimal on overall system performance. As such, it would be ideal if it is implemented in hardware. Since DIM's code transformation is transparent to the user, no modification to executable code is necessary thus leaving the software development process unaffected. Figure 7.22 shows how DIM is integrated in the overall hardware architecture, which consists of a processor coupled with a coarse-grain reconfigurable array while Figure 7.23 shows a configuration example of the configurable array based on the executable code shown in Figure 7.23.(a).

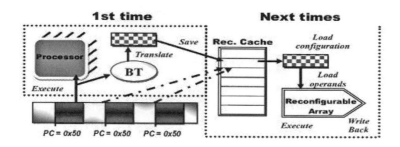

Fig. 7.22   DIM place in system architecture.

The array has a two-dimensional structure where each column contains a number of functional units such as ALUs, multipliers, and shifters. In addition, the array contains load/store units in a separate set of columns within the array. Contrary to complex arithmetic operations, more than one simple operation can be executed in a single processor cycle by the array. However, the array does not support floating point operations. When an instruction is loaded on the array, it is allocated to a given row in the array at a specific column. Instructions without any data dependencies can be allocated on the same row. Input operands are loaded on the array through a set of busses where each bus is connected to a functional unit through two multiplexers.

As Figure 7.22 shows, DIM takes place simultaneously with instruction fetching by the processor. The DIM algorithm takes advantage of the observation that sequences of instructions with the same operands will be repeated a number of times during program execution [Gonzales (1999)]. Detection of these instructions starts with the first instruction encountered after a branch and stops when the next branch instruction is encountered. While the detection phase collects these instructions, they are translated and stored in a special buffer. If the detection phase collects more than three instructions, the translated instructions are saved in a special cache. Each entry in this cache keeps information about the routing of the instruction operands and the configuration bits of the target functional unit in the reconfigurable array. For each instruction, the algorithm resolves read-after-write dependencies by examining the source operands followed by a configuration of the operands on special registers in the configurable array. The DIM algorithm can take advantage of the

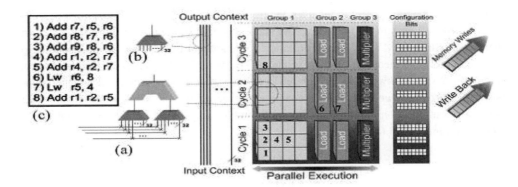

Fig. 7.23   A configuration example of the array

Table 7.5   Hardware configurations used in experimentation

	C #1	C #2	C #3
#Lines	24	48	150
#Columns	11	16	20
#ALU / line	8	8	12
#Multipliers / line	1	2	2
#Ld/St / line	2	6	6

different delays of the functional units of the array. In addition, it can handle false data dependencies and performs speculative execution. The configuration phase of the algorithm starts by loading the configuration bits of the multiplexers and functional units into the array followed by the retrieval of the operands from the cache in order to route them to the functional units in the array. The configuration process can consume three cycles before the execution of the array is activated in the fourth cycle. Occasionally, the configuration process will take more than three cycles if a number of operands need to be fetched from the input registers. In this case, the processor stalls and waits for the reconfiguration step to complete before resuming its normal execution. To evaluate DIM's performance, three configurations are used in experimentation as shown in Table 7.5 where each array configuration is evaluated with 16, 64, and 512-slot caches with speculation up to three basic blocks.

A mix of data and control flow benchmark algorithms were used to test DIM. The obtained speedup results are shown in Table 7.6 where the two rightmost columns labeled ideal represent the anticipated speedups if there are no limits on the reconfigurable resources of the array.

The table shows that it is possible to approach theoretical speedups for most benchmarks

Table 7.6 Obtained speedups of the benchmarks over the three hardware configurations

Algorithm	Speed Up Configuration #1						Speed Up Configuration #2						Speed Up Configuration #3						Ideal	
	No Speculation			Speculation			No Speculation			Speculation			No Speculation			Speculation			No Spec	Spec
	16	64	256	16	64	256	16	64	256	16	64	256	16	64	256	16	64	256		
Rijndael E.	1.05	1.20	1.21	1.05	1.24	1.24	1.06	1.71	1.70	1.06	1.96	1.95	1.06	1.46	1.66	1.06	1.69	1.69	5.18	8.05
Rijndael D.	1.07	1.21	1.21	1.07	1.25	1.25	1.07	1.93	1.94	1.07	1.96	1.95	1.07	3.30	3.95	1.07	2.32	2.32	4.88	7.40
GSM E.	1.93	1.65	1.69	2.01	2.05	2.13	1.53	1.85	1.98	2.03	2.07	2.17	1.63	1.66	1.66	2.03	2.07	2.19	1.70	2.19
JPEG E.	1.93	2.04	2.07	1.79	1.88	1.89	2.50	2.72	2.73	3.55	4.27	4.37	2.50	2.72	2.73	3.55	4.27	4.37	3.33	2.64
SHA	1.90	1.90	1.90	3.91	3.86	3.84	1.90	1.91	1.91	4.90	4.84	4.90	1.90	1.91	1.91	4.80	4.84	4.84	1.91	4.87
Susan Smoothing	1.49	1.50	1.65	2.70	2.99	3.31	1.49	1.61	1.65	2.93	3.14	3.52	1.48	1.61	1.65	2.93	3.14	3.52	1.85	3.52
CRC	1.53	1.53	1.53	1.92	1.92	1.92	1.53	1.53	1.53	1.92	1.92	1.92	1.53	1.53	1.52	1.90	1.92	1.92	1.53	1.92
JPEG D.	1.92	2.03	2.04	1.64	1.78	1.78	2.05	2.21	2.22	2.02	2.54	2.55	2.05	2.21	2.22	2.03	2.62	2.63	2.77	4.36
Patricia	1.49	1.84	1.83	1.58	2.05	2.23	1.49	1.86	1.95	1.95	2.17	2.37	1.45	1.86	1.95	1.64	2.17	2.37	2.18	3.07
Susan Corners	1.22	1.49	1.72	1.31	1.47	1.91	1.38	1.70	2.17	1.56	1.76	2.64	1.38	1.76	2.17	1.56	1.79	2.64	2.37	2.66
Susan Edges	1.23	1.42	1.64	1.29	1.48	1.83	1.43	1.70	2.20	1.47	1.74	2.43	1.43	1.76	2.20	1.53	1.81	2.58	2.21	2.60
Dijkstra	1.59	1.71	1.71	2.03	2.21	2.32	1.59	1.72	1.72	2.04	2.24	2.24	1.58	1.77	1.72	2.04	2.24	2.24	1.72	2.25
GSM D.	1.28	1.28	1.29	1.27	1.28	1.29	1.62	1.62	1.65	1.45	1.50	1.52	2.76	2.79	2.80	2.37	2.49	2.58	3.31	3.86
Blowfish	1.75	1.75	1.76	1.83	1.83	1.82	1.76	1.76	1.76	1.83	1.83	1.83	1.76	1.76	1.76	1.83	1.83	1.83	1.76	1.83
Stringsearch	1.38	1.81	1.88	1.56	2.22	2.77	1.38	1.82	1.89	1.57	2.90	2.67	1.38	1.82	1.88	1.57	2.58	2.94	1.85	2.97
Quicksort	1.37	1.74	1.74	1.89	2.32	2.93	1.37	1.77	1.77	1.80	2.06	2.67	1.37	1.77	1.77	1.88	2.58	2.92	1.77	2.87
RawAudio E.	1.02	1.61	1.61	1.98	1.99	2.00	1.60	1.61	1.61	1.98	1.99	2.00	1.60	1.61	1.61	1.98	1.98	2.00	1.61	2.00
RawAudio D.	1.64	1.64	1.64	1.79	1.79	1.79	1.64	1.64	1.64	1.79	1.79	1.79	1.64	1.64	1.64	1.79	1.79	1.79	1.64	1.79
Average	1.51	1.62	1.68	1.80	1.98	2.09	1.58	1.78	1.86	2.03	2.37	2.49	1.65	2.04	2.13	2.01	2.50	2.67	2.32	3.38

especially dataflow algorithms. In addition, energy consumption is evaluated for the configuration array coupled with the MIPS processor vs. the MIPS processor alone as shown in Figure 7.24. The figure shows that energy savings are significant when coupling the MIPS processor with the array. These savings are obtained through a reduction in the total number of cycles to execute the benchmarks when coupling the MIPS processor with the array. The reduction in the overall number of execution cycles is made possible with the repeated execution of sets of instructions on the configurable array since they reside in its cache without resorting to repeated fetching of the same instructions from the memory. Beside speedup and energy consumption, area cost is evaluated as shown in Table 7.7.

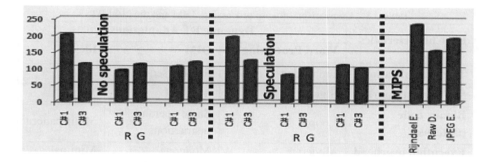

Fig. 7.24 Energy consumption using a 64-slot cache with or without speculation

Table 7.7.(a) shows the number of functional units needed to implement configuration 1 of the array shown in Table 7.3 while Table 7.7.(b) shows the number of bits needed to store one configuration in the array's cache. Finally, Table 7.7.(c) shows the byte capacity of the cache slots used in the array. In terms of gates, the array is equivalent in size to a MIPS R10000 core [Yeager (1996)]. This moderate area cost can be tolerable considering the benefits in terms of speedup and energy savings. These results show that DIM is an excellent example of runtime software adaptation. However, these advantages are made possible primarily by the fast reconfiguration times produced by the coarse-grain array used to sup-

Table 7.7   Area cost of the configurable array

Unit	#	Gates
ALU	192	300,288
LD/ST	36	1,968
Multiplier	6	40,134
Input Mux	408	261,936
Output Mux	216	58,752
DIM Hardware		1,024
Total		664,102

Table	#bits	#Slots	#Bytes
Write Bitman Table	256	2	833
Resource Table	786	4	1,601
Reads Table	1,632	8	3,300
Writes Table	576	16	6,404
Context Start	40	32	13,012
Context Current	40	64	25,616
Immediate Table	128	128	51,304
Total	3,202	256	102,464

port DIM. DIM would have been widely applicable if the reconfigurable hardware was an FPGA instead since FPGAs offer a high degree of customization of the reconfigurable hardware architecture to suit the running application. In brief, DIM expands adaptability at the software level while its configurable array limits it at the hardware level.

7.4.3   Summary of Software Adaptation

Table 7.8 presents a summarized comparison between warp processing, DIM, and the ideal binary translation (BT) adaptation.

Table 7.8   Comparison of warp processing, DIM and the ideal BT

BT Characteristics	Dynamic BT Adaptations		
	Warp Processing	DIM	Ideal
Target applications	Embedded	Parallel/sequential	Any
Software support	Significant	Minimal	None
Binary code patterns	Loops	Same operand instructions	Multi-pattern
Execution overhead	Moderate	Moderate	Minimal
Transparency	Complete	Complete	Complete
Reconfiguration overhead	Significant	Minimal	None
Implementation	Significant hardware	Minimal hardware	Minimal
Reconfigurable hardware	W-FPGA	CGRA	FPGA

Column 1 represents the characteristic of BT that are used to compare the three BT adaptations while column 4 represents the ideal approach in runtime BT. Row 1 of the table shows the applications targeted by the three approaches while row 3 shows the level of support that needs to be provided by software tools for each approach. Row 3 shows the type of patterns within the executable binary code that each approach uses for BT while row 4 shows the overhead added to runtime execution to support each approach. In binary code, several patterns emerge depending on the application. For example, some applications are rich in loops while others exhibit repeated large blocks of code made up of identical instructions that execute repeatedly on different sets of operands. These binary code patterns can be monitored by a BT approach in order to identify which part of the binary code ought to be migrated to reconfigurable hardware. Row 5 shows the degree of transparency displayed by each approach while row 6 shows the overhead required to reconfigure the reconfigurable hardware to accelerate the code identified by each BT approach. Row 7 shows whether the approach is implemented in hardware or software while row 8 shows the type of hardware that each approach is using for accelerating binary code. Regarding hardware implementation, a BT approach can be implemented as a software task running on a processor or embedded in reconfigurable hardware if its area cost is minimal. Of course, there are pros and cons to each type of implementation. However, whichever implementation is capable of maintaining a minimal overhead and complete transparency would be acceptable.

7.5 Future Directions in Runtime Adaptation

The last two sections presented an overview of the potential of adaptation in terms of hardware and software. To extend this vision of adaptation, it would be helpful to formulate the ideal adaptation in terms of hardware and software.

7.5.1 *Future Hardware Adaptation*

For runtime adaptation, the ideal hardware would be the hardware, including implementation technology and architectural innovation, which provides the highest degree of versatility and performance to support the varying demands of an application. Although it is not always possible to achieve a high level of both at the same time, versatility and performance can be realized to some extent with careful choices of implementation hardware technology. However, there is a cost associated with versatility since a high degree of the latter imposes a high overhead in terms of reconfiguration. Figure 26 shows how performance, versatility, and reconfiguration overhead relate to each other in different implementation technologies. While CGRAs can provide excellent performances for target applications, they are not sufficiently versatile to support the performance requirements when the target application changes. CGRAs have been optimized to support a specific classes of target applications. As a result, their reconfigurable resources consist mostly of coarse-grain components, which impose a relatively low reconfiguration overhead. At the other end of the spectrum, FPGAs are the only devices capable of providing both versatility and performance. Their high degree of versatility is possible due to the rich mix of coarse- and fine-grain resources available in most SOC FPGAs such as Virtex chips. In FPGAs, maximum performance can be achieved by spatially mapping computations preferably on fine-grain reconfigurable resources at the expense of a large reconfigurable overhead. However, this reconfiguration overhead can be overcome by adopting instead multi-level reconfigurable architectures. QUKU is an example of a multi-level reconfiguration architecture. In general, an architecture can be reconfigured at bit-word-, or instruction-level. Low Reconfiguration Overhead High FPGA [Lysecky (2006)] Processor [Lysecky (2006)] Montium CGRA [Heysters (2006)] Low Performance & Versatility High Z

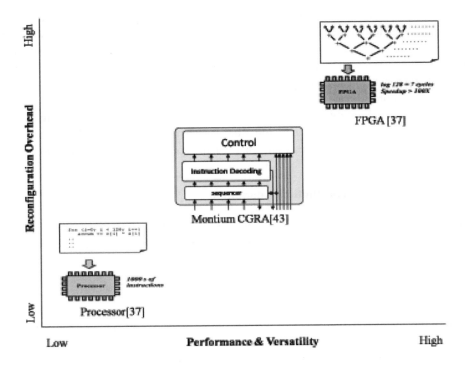

Fig. 7.25   Performance and versatility vs. reconfiguration overhead

These three levels do not by any means subsume other levels of reconfiguration. It is conceivable to formulate architectures which have a higher number of reconfiguration levels. These reconfiguration levels can be thought of as a hierarchical approach using different granularities in reconfiguration. The three reconfiguration levels impose different reconfiguration overhead where this overhead is high at the lowest bit level and low at the highest instruction level. It would be even desirable in future reconfigurable SOCs to have a built-in reconfiguration architecture with different levels of granularity. Currently, commercial FPGAs have yet to offer such advanced reconfiguration capabilities. Table 7.9 shows the overhead, performance, frequency and appropriate usage for each reconfiguration level.

Because bit-level reconfiguration can degrade performance, it would be wise to use it spar-

Table 7.9   Overhead, impact on performance, frequency and usage for the three reconfiguration levels

Parameters	Reconfiguration levels		
	Bit	**Word**	**Instruction**
Overhead	Highest	Moderately low	Lowest
Performance	Hinders	Relatively helps	Helps
Frequency	Lowest	Moderately high	Highest
Usage	Architectural change	Customization	Operation

ingly during runtime. For example, bit-level reconfiguration can be used only to swap in
a new architecture to support a new application which is about to start executing. On the
other hand, word-level reconfiguration does not hamper performance as much as bit-level
reconfiguration does. This reconfiguration can be used to customize dynamically the architecture in order to meet the demands of the application. This customization can be used
for instance to alter the bitwidth of a processor or the datapath of a PE. Finally, because
instruction-level reconfiguration imposes the lowest overhead, it ought to be used more
frequently as in the case of operating a processor on a cycle basis in a multiprocessor architecture. Given the primacy of versatility as a necessary requirement for runtime adaptation,
an ideal architecture would be an architecture that (i) mixes and matches its components
with the variety of reconfigurable resources available in SOC FPGAs, and (ii) minimizes
the reconfiguration overhead as much as possible by devising architectural features to do
so.

7.5.2   *Future Software Adaptation*

For runtime adaptation, the ideal approach in BT would be the one shown in the rightmost column of Table 7.7 and labeled as Ideal. This approach does not need any support
from software tools and require any modification to the compilation process. This has the
advantage of running a variety of binary code including legacy applications. Regardless
of the target application, binary code tends to exhibit various repetitive patterns in its instruction content. These patterns can be viewed as similar design patterns found in the
software engineering world [Gamma (1995)]. It is highly desirable if most repetitive patterns in binary code can be studied and identified for the purpose of improving future BT
approaches. Although the research literature related to the issue of patterns identification
in binary code is not exhaustive at this point, past efforts can help in assembling a limited
catalog of binary code patterns [Gonzales (1999); Gschwing (2000)]. While current BT
approaches focus on a single pattern in binary code, future BT approaches ought to expand
their focus to detect and identify a variety of binary code patterns based on the assembled
catalog of patterns. This will make these approaches widely applicable to a large number of applications. However, a BT scheme that identifies and detects a large number of

patterns may require an implementation that may not be minimal regardless of whether the implementation of this approach is in the software or hardware domain. To maintain the non-intrusiveness nature of these BT approaches, a level of reconfiguration either on the software or hardware implementation, is necessary and must be integrated in the design anticipated for these approaches. BT reconfiguration can be reached by adopting a modular design where each module is responsible for the identification and detection of a small number of patterns. Modules can be activated only when needed. By maintaining non-intrusion through a modular design, execution overhead can be minimized while transparency can be guaranteed. Of course, these new BT approaches must somehow be tailored to migrate their instruction streams across different resources within SOC FPGAs in order to take advantage of the high degree of versatility provided by these chips. In that case, the translation process in these BT approaches must take advantage of the multi-level nature of reconfiguration that can be provided by an optimized architecture mapped on the SOC. From this perspective, there must be a well-planned tight coupling between a BT approach and an adaptive architecture mapped onto an SOC FPGA. Ample evidence is available to show that this coupling may be challenging due to the awkward nature of the reconfiguration infrastructure offered by current SOC FPGAs. Reconfiguration infrastructures that can accommodate fast runtime adaptation are necessary in order to bridge the gap between software and hardware adaptation.

7.5.3 *Future Reconfiguration Infrastructures in FPGA SOCs*

To exploit the ultimate power of runtime adaptation, it is important if the adaptation capabilities on the software and hardware sides are cleverly integrated from a system-level perspective. This would provide the computing resources needed to support multimedia-centric applications. However, the great potential of runtime adaptation is not duly served with the inadequate reconfiguration capabilities provided by current FPGAs. Today's FPGAs are still unable to provide reconfiguration times in the nanosecond range especially when the reconfiguration bitstreams are quite large. To overcome these reconfiguration shortcomings in FPGAs, several efforts have been proposed in the past. One approach has already been mentioned in this chapter, namely the use of multi-level reconfiguration architectures such as the QUKU architecture [Shukla (2006)]. In this approach, word-level reconfiguration must be used as frequently as possible while bit-level reconfiguration must be used as little as possible. The former can be performed relatively fast as opposed to the latter. Other efforts focused on compressing the configuration bitstream at design time and decompressing it at run time [Pan (2004); Li (2001); Dandalis (2005)]. Furthermore, others proposed cloning a configuration by taking advantage of regularity and locality in the configuration bitstream. This approach is used to copy configuration bits from one area in the reconfigurable fabric of the FPGA to another in order to reduce configuration latency [Park (1999)]. While these techniques can reduce configuration download time, they are not sufficiently effective to provide a powerful solution to the integration of software and hardware runtime adaptation. To confront this challenge, a complete rethinking of the reconfiguration infrastructure in FPGAs is necessary. In this context, it is meant by recon-figuration infrastructure the actual hardware architecture used in current FPGAs to upload a given configuration from the outside or over-write from inside the FPGA chip. Most FPGA configuration architectures do not provide sufficiently high bandwidth needed for runtime adaptation. This problem is even severe in large capacity FPGAs where writing megabyte streams to the chip is possible. A possible approach to address this problem is to view configuration data as any other data used in an FPGA chip. This data can be application-(e.g. operands), program- (e.g., instructions), or configuration-related data (e.g., bitstream). Executing code or modifying architectural elements is reduced in essence to moving data (from outside or inside the chip) in the form of bits to the proper reconfigurable resource inside the FPGA chip. However, moving data requires communication backbones such as busses or networks. On one hand, current FPGA chips offer a visible network of routing re-sources (e.g., wires, switches, etc.) for moving data on the reconfigurable logic subsystem. On the other hand, these chips offer cumbersome access to a somewhat invisible network of wires and registers to move and store configuration data on another configuration sub-system. It is possible to overhaul these two subsystems by merging them into a single bandwidth-powerful system that moves data efficiently regardless of its nature. Recently, many efforts began to point to this direction by extolling the advantages of using NOCs on FPGAs to overcome a number of costly implementation factors. Until recently, most NOC designs were proposed as soft structures that can be mapped on the reconfigurable fabric of the chips. While these softly mapped NOCs can be highly flexible, they tend to consume a non-trivial amount of resources that could be used for logic instead of routing. The alterna-tive would be to hardwire these NOCs on the chip itself in order to route "packets and not wires" at high speed over long distances with minimal power consumption [Hect (2005); Goossens (2008)]. A possible concept of a hardwired NOC is shown in Figure 27 where the tiles represent possible locations for reconfigurable logic. Because most current FPGA designs resemble large SOCs rich in busses and FIFOs, it would not be unreasonable to hardwire an NOC on an FPGA chip. By doing so, the same NOC can be used to support two planes where the first plane is dedicated to data-transfer messages and protocols while the second plane is dedicated to move configuration data. At this point, it is not clear if a hardwired NOC can deliver its promise. If that is the case, it is not clear what architec-ture this NOC should have and what communication protocols it should support. Issues regarding hardwired NOCs remain not well understood.

7.6 Conclusion

This chapter presents the technical challenges associated with meeting the performance requirements of new multimedia-centric applications. It discusses how current software and hardware approaches have reached an inflection point beyond which any substantial benefits for boosting the performance of these applications are difficult to obtain. The chapter makes an argument for runtime adaptation in the software and hardware domain as

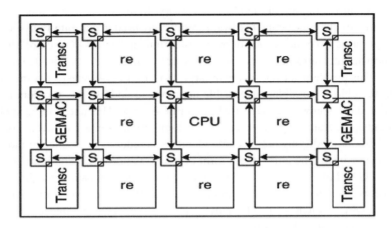

Fig. 7.26   Diagram of a hardwired NOC on an FPGA SOC

a possible direction for overcoming these stringent performance requirements. To this end, detailed explanations of examples of promising hardware and software adaptation are presented. These examples are used to characterize the ideal hardware and software runtime adaptation. The only obstacle to runtime adaptation is the bandwidth-limiting reconfiguration infrastructures available in most FPGA SOCs. The chapter puts forward a proposal for a complete rethinking of reconfiguration by suggesting hardwired NOCs on FPGA SOCs for both communication and reconfiguration. In the overall, system designers and architects will be faced with the following pressing questions for some time: Given a class of embedded applications, is it possible to develop better runtime software adaptation techniques which can sustain the performance of the application? Performance parameters vary among applications (e.g., power, throughput, user response, etc.). Given a class of embedded applications, is it possible to develop better runtime hardware adaptation techniques to accommodate the workload variations of the application? Are there clever approaches to unify runtime adaptation in the software and hardware domain within a single system architecture that can be applied to a class of applications? How can the reconfiguration bottleneck be overcome considering the unification of software and hardware adaptation from a system perspective? Using the answers from the above question, can all or some of these techniques be extended to other performance-demanding applications such as scientific computing and virtual reality? Answering these questions can provide much needed breakthroughs to help serve user demands for new usage modalities of information technology.

Bibliography

T. Bjerregaard, S. Mahadevan, A survey of research and practices of Network-on-chip, ACM Computing Surveys (CSUR), Volume 38, Issue 1, (2006).

K. Lee, S. J. Lee, H. J. Yoo, Low-power network-on-chip for high-performance SoC design, IEEE Transactions on Very Large Scale Integration (VLSI) Systems, Volume 14, Issue 2, Feb. 2006, pp. 148–160.

S. J. Lee, K. Lee; H. J. Yoo, Packet -Switched On-chip Interconnection Network for System-on-Chip Applications, IEEE Trans. on Circuit and Systems, Vol. 52, No. 6, June 2005.

P.P. Pande *et al.*, "Performance Evaluation and Design Trade-offs for Network-on-Chip Interconnect Architectures," IEEE Trans. Computers, Vol. 54, No. 8, Aug. 2005, pp. 1025–1040.

D. Bertozzi, A. Jalabert *et al.*, NoC Synthesis Flow for Customized Domain Specific Multiprocessor Systems-on-chip. IEEE Transaction on Parallel and Distributed Systems, Vol. 16, No.2 Feb, 2005.

L. Benini *et al.*, "Networks on chips: A new SoC paradigm, EIEEE Computer, Vol. 36,No. 1, pp. 70E8, Jan. 2002.

W.J. Dally, B. Towles, Route packets, not wires: on-chip interconnection networks, DAC 2001, June 18E2, 2001, Las Vegas, Nevada, USA.

P. Guerrier, A. Greiner, A generic architecture for on-chip packet-switched interconnections, in: Design, Automation and Test in Europe Conference and Exhibition 2000, Proceedings, 2000, pp. 250E56.

E. Rijpkema, K. Goosens, P. Wielage, A router architecture for networks on silicon, Proceedings of Progress 2001, 2nd Workshop on Embedded Systems.

S. Kumar, A. Jantsch, J.-P. Soininen, M. Forsell, M. Millberg, J. Oberg, K. Tiensyrja, A. Hemani, A network on chip architecture and design methodology, Proceedings of the IEEE Computer Society Annual Symposium on VLSI, 2002 (ISVLSI.02).

K. Fall and K. Varadhan, "The ns ManualE pp.1-380, Dec.2003.

L. Benini and D. Bertozzi, "Xpipes: A Network-on-Chip Architecture for Gigascale Systems-on-Chip, IEEE Circuits and Systems Magazine, Vol. 4, No. 2, Apr.-June, 2004, pp. 18-31.

P.P. Pande, C. Grecu, A. Ivanov, and R. Saleh, "High-Throughput Switch-Based Interconnect for Future SoCs, E Proc. Third IEEE Int'l Workshop System-on-Chip for Real-Time Applications pp. 304-310, 2003.

W.J. Dally and C.L. Seitz, "Deadlock-Free Message Routing in Multiprocessor Interconnection Networks, "IEEE Trans. Computers, Vol. C-36, No. 5, May 1987, pp. 547-553.

W. J. Dally, "Virtual-channel flow control", Parallel and Distributed Systems, IEEE Transactions on Volume 3, Issue 2, March 1992, pp.194-205.

ITRS. 2003. International technology roadmap for semiconductors. Tech. rep., International Technology Roadmap for Semiconductors.

L. M. Ni and P. K. McKinley, "A Survey of Wormhole Routing Techniques in Direct Networks, " IEEE Comp.Mag., vor 26, No. 2, Feb. 1993, pp. 62-76.

D. Sylvester and K. Keutzer. A Global Wiring Paradigm for Deep Submicron DesignE IEEE Transactions on Computer Aided Design of Integrated Circuits and Systems, pages 242E52, February 2000.

J. Davis and D. Meindl. Compact Distributed RLC Interconnect Models - Part II: Coupled Line Transient Expressions and Peak Crosstalk in Multilevel NetworksE IEEE Transactions on Electron Devices, 47(11):2078E087, November 2000.

A. Radulescu, J. Dielissen *et al.*, An efficient on-chip NI offering guaranteed services, shared-memory abstraction, and flexible network configuration, pp. 4-17, IEEE transaction on CAD of Integrated Circuits and Systems, Vol. 24, No1., 2004.

BANC: Basic Network on Chips Project: http://www2.sowa.is.uec.ac.jp/ ben/espoir/

M. T. Rose, The Open Book: A Practical Perspective on OSI. Englewood Cliffs, NJ: Prentice-Hall, 1990.

D. Sylrcstrr, C. hl. Ho *et al.* "htcrcomect Scaling: Signal Iutsgrify aid Perfommnce in Future High-sped CMOS Designs". Io: Proc. of VLSI Symposium 011 Technology, pp. 42-43. 1998.

R. R. Harrison, "A low-power, low-noise CMOS amplifier for neural recording applications, in Proc. IEEE Int. Symp. Circuits and Systems, Vol. 5, 2002, pp. 197-200.

E. Dupont, M. Nicolaidis, and P. Rohr, "Embedded Robustness IPs for Transient-Error-Free ICsE IEEE Design and Test of Computers, Vol. 19, No 3, pp. 56-70, May-June, 2002.

A. Ben-Abdallah, T. Yoshinaga, M. Sowa, High-Level Modeling and FPGA Prototyping of Produced Order Parallel Queue Processor Core, Journal of supercomputing, Vol. 38, Number 1 / October, pp. 3-15, 2006.

D. Flynn, "AMBA: enabling reusable on-chip designs", IEEE Micro, Vol. 17, n. 4, July 1997, pp. 20-27.

IBM CoreConnect Bus Architecture, www-03.ibm.com/chips/products/coreconnect/

D. Wingard and A. Kurosawa, "Integration Architecture for System-on-a-Chip DesignE Proceedings of IEEE 1998 Custom integrated Circuits Conference, May 1998, pp. 85-88.

G. De Micheli, R. Ernst and W. Wolf, *"Readings in Hardware/Software co-design"*, Morka Kaufmann Publishers, ISBN 1-55860-702-1.

M. Sowa, A. Ben-Abdallah and T. Yoshinaga, *"Parallel Queue Processor Architecture Based on Produced Order Computation Model"*, Int. Journal of Supercomputing, Vol.32, No.3, June 2005, pp.217-229.

A. Ben-Abdallah, M. Arsenji, S. Shigeta, T. Yoshinaga and M. Sowa, *"Queue Processor for Novel Queue Computing Paradigm Based on Produced Order Scheme"*, Proc. of HPC, IEEE CS, Jul. 2004, pp. 169-177.

S. Aditya, B. R. Rau and V. Kathail, *"Automatic Architectural Synthesis of VLIW and EPIC Processors"*, Proc. 12th Int. Symposium of System Synthesis, IEEE CS Press, Los Alamitos, Calif., 1999, pp. 107-113.

M. Sheliga and E. H. Sha, *"Hardware/Software Co-design With the HMS Framework"*, Journal of VLSI Signal Processing Systems, Vol. 13, No.1, 1996, pp. 37-56.

S. Chaudhuiri, S. A. Btlythe and R. A. Walker, *"A solution methodology for exact design space exploration in a three dimensional design space"*, IEEE Transactions on VLSI Systems, Vol. 5, 1997, pp.69-81.

D. Lewis, V. Betz, D. Jefferson *et al.*, *"The Stratix Routing and Logic Architecture"*, in Proc. IEEE FPGAs, Monterey, CA, 2003, pp. 12E0.

Cadence Design Systems:http://www.cadence.com/.

Altera Design Software: http://www.altera.com/.

IEEE Standard for Binary Floating-point Arithmetic, ANSI/IEEE Standard 754, 1985.

IEEE task P754, "A proposed standard fr binary floating-ponit arithmetic", IEEE Computer, Vol. 14, No. 12, pp.51-62, March 1981.

Mirko Loghi, Federico Angiolini, Davide Bertozzi, Luca Benini, Roberto Zafalon, "Analyzing On-Chip Communication in a MPSoC Environment, Proceedings of the conference on Design", Design Automation and test in Europe, Vol.2, Feb.16-20, 2004.

R. Ernst, J. Henkel, T. Benner, "Hardware-software co synthesis for microcontrollers", IEEE Design and Test, Dec. 1993, pp. 64-75.

A. D. Booth, "A signed binary multiplication technique, EQuart. J. Mech. Appl. Math., Vol. 4, 1951, pp. 23–40.

A. Jerraya , Multiprocessor System-on-Chip, Morgan Kaufman Publishers, ISBN:0-12385-251-X, 2005.

S. Prakash and A. Parker, "SoS: Synthesis of application-specific heterogeneous multiprocessor systems", J. Parallel Distributed Computing, Vol. 16, 1992, pp. 338–351.

B. Dave, G. Lakshminarayama, and N. Jha, "COSFA: Hardware-software co-synthesis of heterogeneous distributed embedded system architectures for low overhead fault tolerance", In Proc. IEEE fault-Tolerant computing symposium, 1997, pp. 339–348.

C. K. Lennard, P. Schaumont, G. de Jong, A. Haverinen, and P. Hardee, "Standards for System-Level Design: Practical Reality or Solution in Search of a Question?", Proc. Design Automation and Test in Europe, Mar. 2000, pp. 576–585.

A. Ben-Abdallah, Sotaro Kawata, Tsutomu Yoshinaga, and Masahiro Sowa, "Modular Design Structure and High-Level Prototyping for Novel Embedded Processor Core", Proceedings of the 2005 IFIP International Conference on Embedded And Ubiquitous Computing (EUC'2005), Nagasaki, Japan, Dec. 6-9, 2005, pp. 340–349.

S. Pasricha, N. Dutt, M. Ben-Romdhane, "Constraint-Driven Bus Matrix Synthesis for MPSoC", Asia and South Pacific Design Automation Conference (ASPDAC 2006), Yokohama, Japan, January 2006, pp. 30–35.

J. Kistler, Disconnected operation in a distributed file system, PhD thesis, Carnegie Mellon University, School of Computer Science, (1993).

J.R. Lorch and A.J. Smith, Software strategies for portable computer energy management, IEEE Personal Communications, 5 (3) 60E3, (1998).

H. Balakrishnan, V. Enkata, N. Padmanabhan, A Comparison of Mechanisms for Improving TCP Performance over Wireless Links, IEEE/ACM Transaction on Networking, 5 (6), pp. 756–769, (1997).

P. J. Havinga, G. Smit, M. Bos M., Energy-efficient wireless ATM design, proceedings wmATME9, June 2-4, (1999).

F. Akyildiz, S. Weilian , S. Yogesh, and E. Cayirci, A Survey on Sensor Networks, IEEE Communications Magazine, pp. 102–114, (2002).

K.M. Sivalingam, J.C. Chen, P. Agrawal and M. Srivastava, Design and analysis of low-power access protocols for wireless and mobile ATM networks, Wireless Networks 6 (1), 73E7, (2000) .

J. Liang et al., An architecture and compiler for scalable on-chip communication, IEEE Trans. on VLSI Systems, 12 (7), (2004).

V. Tiwari, S. Malik, and A. Wolfe. Power analysis of embedded software: A first step towards software power minimization, IEEE Transactions on Very Large Scale Integration, 2 (4):437E45, (1994).

Intel Corporation. Mobile Power Guidelines 2000. ftp://download.intel.com/design/mobile/ intelpower/mpg99r1.pdf, December 1998.

F. Wolf, Behavioral Intervals in Embedded Software: Timing and Power Analysis of Embedded Real-Time Software Process, Kluwer Academic Publishers, ISBN 1-4020-7135-3, 2002

J. Lorch, A complete picture of the energy consumption of a portable computer. Master's thesis, Department of Computer Science, University of California at Berkeley, (1995).

T. Martin, Balancing batteries, power and performance: System issues in CPU speed-setting for mobile computing, Ph.D. Dissertation, Carnegie Mellon University, Department of Electrical and Computer Engineering, Aug. (1999).

Gregory F. Welch, A Survey of Power Management Techniques in Mobile Computing Operating Systems, ACM SIGOPS Operating Systems Review, Volume 29, Issue 4, (1995).

R. Kravets and P. Krishnan, Application driven power management for mobile communication Springer Science, Wireless Networks, 6 (4), pp. 263–277, (2000).

J.M. Rulnick and N. Bambos, Mobile power management for maximum battery life in wireless communication networks, in: Proceedings of IEEE INFOCOM E6, (1996).

L. Benini and G. de Micheli, System-level power optimization: Techniques and tools, in Proc. Int. Symp. Low-Power Electronics Design, San Diego, CA, pp. 288E93, (1999).

Intel Corporation. Mobile Intel Pentium III processor in BGA2 and micro-PGA2 packages, revision 7.0, (2001).

J. R. Lorch and A. J. Smith. Reducing processor power consumption by improving processor time management in a single user operating system. In Second ACM International Conference on Mobile Computing and Networking (MOBICOM), (1996).

M. Weiser, B. Welch, A. Demers, and S. Shenker. Schedlibng for reduced cpu energy. In Proceedings of the First Symposium on Operating System Design and Implementation (OSDI) E4,(1994).

K. Govil, E. Chan, and H. Wasserman. Comparing algorithms for dynamic speed-setting of a low-power cpu, In First ACM International Conference on Mobile Computing and Networking (MOBICOM), (1995).

C.L. Su and Alvin M. Despain, Cache Designs for Energy Efficiency, in Proc. of the 28th Hawaii International Conference on System Science, (1995).

C. Su and A. Despain, Cache Design Tradeoffs for Power and Performance Optimization: A Case Study, Proc. of International Symposium on Low Power Design, (1995).

P. Erik, P. Harris, W. Steven , E. W. Pence, S. Kirkpatrick, Technology directions for portable computers. Proceedings of the IEEE, 83b(4): 63–57, (1995).

F. Doughs, P. Krishnan, and B. Marsh, Thwarting the power hungry disk. In Proceedings of the 1991 Winter USENIX Conference, (1994).

K. Li, R. Kumpf, P. Horton, and T. Anderson, A quantitative analysis of disk drive power manage ment in portable computers. In Proceedings of the 1994 Winter USENIX, (1994).

F. Douglis, F. Kaashoek, B. March, R. Caceres, K. Li, and J. Tauber, storage alternative for mobile computers, Proceedings of the first USENIX Symposimum on Operating Systems Design and IMplemnetation, (1994).

P. J. M. Havinga and G. J. M. Smit, Energy-efficient wireless networking for multimedia applications, in Wireless Communications and Mobile Computing. New York: Wiley, Vol. 1, pp.165E84, (2001).

F. N. Najm, A survey of power estimation techniques in VLSI circuits, IEEE Trans. VLSI Syst., 2 (4), pp. 44–55, (1994).

M. Pedram, Power minimization in IC design: Principles and applications, ACM Trans. Design Automat. Electron. Syst., 1 (1), pp. 3–6, (1996).

Multiprocessor System-on-Chip, Morgan Kaufman Pblishers, ISBN:0-12385-251-X, (2005).

Semiconductor Industry Association: "The national technology rodmap for semiconductors: technology needs", Sematche Inc., htt://www.sematech.org, Austin, USA, (1997).

S. Borkar, Design challenges of technology scaling. IEEE Micro, 19 (4), (1999).

Y. Ye, S. Borkar, and V. De, A New Technique for Standby Leakage Reduction in High-Performance Circuits, 1998 Symposium on VLSI Circuits, pp. 40E1, Honolulu, Hawaii, (1998)

L. Benini, G. de Micheli, and E. Macii, Designing low-power circuits: Practical recipes, IEEE Circuits Syst. Mag., Vol. 1, pp. –5, (2001).

A. Ben-Abdallah, Sotaro Kawata, Tsutomu Yoshinaga, and Masahiro Sowa, Modular Design Structure and High-Level Prototyping for Novel Embedded Processor Core. Proceedings of the 2005 IFIP International Conference on Embedded And Ubiquitous Computing (EUC'2005), Nagasaki, Dec. 6-9, pp. 340–349, (2005).

V. Tiwari, S. Malik, and P. Ashar. Guarded Evaluation: Pushing Power Management to Logic Synthesis/Design. IEEE Transactions on Computer Aided Design of Integrated Circuits and Systems, 17 (10):1051E060, (1998).

A. Abnous, J. Rabaey, Ultra-Low-Power Domain-Specific Multimedia Processors, Proceedings of the IEEE VLSI Signal Processing Workshop, IEEE press, pp. 459-464, (1996).

J. Rabaey, L. Guerra, R.Mehra, Design guidance in the Power Dimension, Proceedings of the ICASSP, (1995).

H. Mehta, R. M. Owens, M. J. Irwin, R. Chen, and D. Ghosh, Techniques for Low Energy Software, in Internatzonal Symposzum of Low Power Electronics and Deszgn, pp. 72-75, IEEE/ACM, (1997).

J. Lorch, Modeling the effect of different processor cycling techniques on power consumption, Performance Evaluation Group Technical Note 179, ATG Integrated Sys., Apple Computer, (1995).

Mike Galles, Scalable Pipelined Interconnect for Distributed Endpoint Routing: The SGI SPIDER Chip, Proceedings of the IEEE Symposium on High-Performance Interconnects (HOTI'96), pp. 141–146, Aug, 1996.

Hiroki Matsutani, Michihiro Koibuchi, Hideharu Amano, and Tsutomu Yoshinaga, Prediction Router: Yet Another Low Latency On-Chip Router Architecture, Proceedings of the International Symposium on High-Performance Computer Architecture (HPCA'09), pp. 367–378, Feb, 2009.

Akbar Sharifi, Reza Sabbaghi-Nadooshan, and Hamid Sarbazi-Azad, The Shuffle-Exchange Mesh Topology for 3D NoCs, Proceedings of the International Symposium on Parallel Architectures, Algorithms, and Networks (I-SPAN'08), pp. 275–280, May, 2008.

Frank Thomson Leighton, New Lower Bound Techniques for VLSI, Mathematical Systems Theory, 17 (1), pp. 47–70, Apr, 1984.

Reza Sabbaghi-Nadooshan, Mehdi Modarressi, and Hamid Sarbazi-Azad, A Novel High-Performance and Low-Power Mesh-Based NoC, Proceedings of the International Workshop on Performance Modeling, Evaluation, and Optimization of Ubiquitous Computing and Networked Systems (PMEO-UCNS'08), Apr, 2008.

Robert Mullins, Andrew West, and Simon Moore, The Design and Implementation of a Low-Latency On-Chip Network, Proceedings of the Asia and South Pacific Design Automation Conference (ASP-DAC'06), pp. 164–169, Jan, 2006.

Dongkook Park, Reetuparna Das, Chrysostomos Nicopoulos, Jongman Kim and Narayanan Vijaykrishnan, Ravishankar Iyer, and Chita Das, Design of a Dynamic Priority-Based Fast Path Architecture for On-Chip Interconnects, Proceedings of the IEEE Symposium on High-Performance Interconnects (HOTI'07), pp. 15–20, Aug, 2007.

C. Izu, R. Beivide, and C. Jesshope, Mad-Postman: A Look-Ahead Message Propagation Method for Static Bidimensional Meshes, Proceedings of the Euromicro Workshop on Parallel and Distributed Processing (PDP'94), pp. 117–124, Jan, 1994.

George Michelogiannakis, Dionisios N. Pnevmatikatos, and Manolis Katevenis, Approaching Ideal NoC Latency with Pre-Configured Routes, Proceedings of the International Symposium on Networks-on-Chip (NOCS'07), pp. 153–162, May, 2007.

Michihiro Koibuchi, Hiroki Matsutani, Hideharu Amano, and Timothy Mark Pinkston, A Lightweight Fault-tolerant Mechanism for Network-on-Chip, Proceedings of the International Symposium on Networks-on-Chip (NOCS'08), pp. 13–22, Apr, 2008.

M. Thottethodi, A. R. Lebeck, and S. S. Mukherjee, BLAM : A High-Performance Routing Algorithm for Virtual Cut-Through Networks, International Parallel and Distributed Processing Symposium, pp. 45b, (2003).

John Kim, William J. Dally, Brian Towles, and Amit K. Gupta, Microarchitecture of a High-radix Router, Proceedings of the International Symposium on Computer Architecture (ISCA'05), pp. 420–431, Jun, 2005.

John Kim, James Balfour, and William J. Dally, Flattened Butterfly Topology for On-Chip Networks, Proceedings of the International Symposium on Microarchitecture (MICRO'07), pp. 172–182, Dec, 2007.

Marcello Coppola, Riccardo Locatelli, Giuseppe Maruccia, Lorenzo Pieralisi, and Alberto Scandurra, Spidergon: a novel on-chip communication network, Proceedings of the International Symposium on System-on-Chip (ISSOC'04), pp. 15, Nov, 2004.

G. Della Vecchia, and C. Sanges, A Recursively Scalable Network VLSI Implementation, Future Generation Computer Systems, 4 (3), pp. 235–243, Oct, 1988.

Dara Rahmati, Abbas Eslami Kiasari, Shaahin Hessabi, and Hamid Sarbazi-Azad, A Performance and Power Analysis of WK-Recursive and Mesh Networks for Network-on-Chips, Proceedings of the International Conference on Computer Design (ICCD'06), pp. 142–147, Oct, 2006.

Amit Kumar, Li-Shiuan Peh, Partha Kundu, and Niraj K. Jha, Express Virtual Channels: Towards the Ideal Interconnection Fabric, Proceedings of the International Symposium on Computer Architecture (ISCA'07), pp. 150–161, Jun, 2007.

Tushar Krishna, Amit Kumar, Patrick Chiang, Mattan Erez, and Li-Shiuan Peh, NoC with Near-Ideal Express Virtual Channels Using Global-Line Communication, Proceedings of the IEEE Symposium on High-Performance Interconnects (HOTI'08), pp. 11–20, Aug, 2008.

Li-Shiuan Peh, and William J. Dally, A Delay Model and Speculative Architecture for Pipelined Routers, Proceedings of the International Symposium on High-Performance Computer Architecture (HPCA'01), pp. 255-266, Jan, 2001.

Jose Flich, A. Mejia, Pedro Lopez, and Jose Duato, Region-Based Routing: An Efficient Routing Mechanism to Tackle Unreliable Hardware in Network on Chips, Proceedings of the International Symposium on Networks-on-Chip (NOCS), pp. 183–194, May, 2007.

Srinivasan Murali, David Atienza, Luca Benini, and Giovanni De Micheli, A multi-path routing strategy with guaranteed in-order packet delivery and fault-tolerance for networks on chip, Proceedings of the Design Automation Conference (DAC), pp. 845–848, Jul, 2006.

Jung-Chun Kao, and Radu Marculescu, Energy-Aware Routing for E-Textile Applications, Proceedings of the Design Automation and Test in Europe (DATE), Vol. 1, pp. 184–189, (2005).

Srinivasan Murali, Theo Theocharides, Narayanan Vijaykrishnan, Mary Jane Irwin, Luca Benini, and Giovanni De Micheli, Analysis of Error Recovery Schemes for Networks on Chips, IEEE Design & Test of Computers, 22 (5), pp. 434–442, (2005).

Ming Li, Qing-An Zeng, and Wen-Ben Jone, DyXY - a proximity congestion-aware deadlock-free dynamic routing method for network on chip, Proceedings of the Design Automation Conference (DAC), pp. 849–852, Jul, 2006.

Evgeny Bolotin, Israel Cidon, Ran Ginosar, and Avinoam Kolodny, Routing Table Minimization for Irregular Mesh NoCs, Proceedings of the Design Automation and Test in Europe (DATE), Apr, 2007.

John L. Henessy, and David A. Patterson, Computer Architecture: A Quantitative Approach Fourth Edition, App E, Morgan Kaufmann, (2006).

Hiroki Matsutani, Michihiro Koibuchi, Yutaka Yamada, Akiya Jouraku, and Hideharu Amano, Non-Minimal Routing Strategy for Application-Specific Networks-on-Chips, Proceedings of the International Conference of Parallel Processing (ICPP'05) Workshops, pp. 273–280, Jun, 2005.

Hiroki Matsutani, Michihiro Koibuchi, and Hideharu Amano, Performance, Cost, and Energy Evaluation of Fat H-Tree:A Cost-Efficient Tree-Based On-Chip Network, Proceedings of the International Parallel and Distributed Processing Symposium (IPDPS'07), Mar, 2007.

Hiroki Matsutani, Michihiro Koibuchi, Daihan Wang, and Hideharu Amano, Run-Time Power Gating of On-Chip Routers Using Look-Ahead Routing, Proceedings of the Asia and South Pacific

Design Automation Conference (ASP-DAC'08), pp. 55–60, Jan, 2008.

Yulu Yang, Akira Funahashi, Akiya Jouraku, Hiroaki Nishi, Hideharu Amano, and Toshinori Sueyoshi, Recursive Diagonal Torus: An Interconnection Network for Massively Parallel Computers, IEEE Transactions on Parallel and Distributed Systems, 12 (7), pp. 701–715, Jul, 2001.

Hiroki Matsutani, Michihiro Koibuchi, and Hideharu Amano, Enforcing Dimension-Order Routing in On-Chip Torus Networks without Virtual Channels, Proceedings of the International Symposium on Parallel and Distributed Processing and Applications (ISPA'06), pp. 207–218, Nov, 2006.

Michihiro Koibuchi, Kenichiro Anjo, Yutaka Yamada, Akiya Jouraku, and Hideharu Amano, A Simple Data Transfer Technique using Local Address for Networks-on-Chips, IEEE Transactions on Parallel and Distributed Systems, 17 (12), pp. 1425–1437, (2006).

Xuning Chen, and Li-Shiuan Peh, Leakage Power Modeling and Optimization in Interconnection Networks, Proceedings of the International Symposium on Low Power Electronics and Design (ISLPED'03), pp. 90–95, Aug, 2003.

Li Shang, Li-Shiuan Peh, and Niraj K. Jha, Dynamic Voltage Scaling with Links for Power Optimization of Interconnection Networks, Proceedings of the International Symposium on High-Performance Computer Architecture (HPCA'03), pp. 79–90, Jan, 2003.

Vassos Soteriou, and Li-Shiuan Peh, Exploring the Design Space of Self-Regulating Power-Aware On/Off Interconnection Networks, IEEE Transactions on Parallel and Distributed Systems, 18 (3), pp. 393–408, Mar, 2007.

Vassos Soteriou, and Li-Shiuan Peh, Design-Space Exploration of Power-Aware On/Off Interconnection Networks, Proceedings of the International Conference on Computer Design (ICCD'04), pp. 510–517, Oct, 2004.

Timothy Mark Pinkston, and Jeonghee Shin, Trends Toward On-Chip Networked Microsystems, International Journal of High Performance Computing and Networking, 3 (1), pp. 3–18, Sep, 2005.

Robert Mullins, Andrew West, and Simon Moore, The Design and Implementation of a Low-Latency On-Chip Network, Proceedings of the Asia and South Pacific Design Automation Conference (ASP-DAC'06), pp. 164–169, Jan, 2006.

Robert Mullins, Andrew West, and Simon Moore, Low-Latency Virtual-Channel Routers for On-Chip Networks, Proceedings of the International Symposium on Computer Architecture (ISCA'04), pp. 188–197, Jun, 2004.

Sriram Vangal, Jason Howard, Gregory Ruhl, Saurabh Dighe, Howard Wilson, James Tschanz, David Finan, Priya Iyer, Arvind Singh, Tiju Jacob, Shailendra Jain, Sriram Venkataraman, Yatin Hoskote, and Nitin Borkar, An 80-Tile 1.28TFLOPS Network-on-Chip in 65nm CMOS, Proceedings of the International Solid-State Circuits Conference (ISSCC'07), pp. 184–185, Feb, 2007.

Masakatsu Nakai, Satoshi Akui, Katsunori Seno, Tetsumasa Meguro, Takahiro Seki, Tetsuo Kondo, Akihiko Hashiguchi, Hirokazu Kawahara, Kazuo Kumano, and Masayuki Shimura, Dynamic Voltage and Frequency Management for a Low-Power Embedded Microprocessor, IEEE Journal of Solid-State Circuits, 40 (1), pp. 28–35, Jan, 2005.

Kevin Nowka, Gary Carpenter, Eric Mac Donald, Hung Ngo, Bishop Brock, Koji Ishii, Tuyet Nguyen, and Jeffrey Burn, A 0.9V to 1.95V Dynamic Voltage-Scalable and Frequency-Scalable 32b PowerPC Processor, Proceedings of the International Solid-State Circuits Conference (ISSCC'02), pp. 340–341, Feb, 2002.

Kentaro Kawakami, Jun Takemura, Mitsuhiko Kuroda, Hiroshi Kawaguchi, and Masahiko Yoshimoto, A 50% Power Reduction in H.264/AVC HDTV Video Decoder LSI by Dynamic Voltage Scaling in Elastic Pipeline, IEICE Transactions on Fundamentals of Electronics, Communications and Computer Sciences, Vol. E89-A, num. 12, pp. 3642–3651, Dec, 2006.

Jeffrey M. Stine, and Nicholas P. Carter, Comparing Adaptive Routing and Dynamic Voltage Scaling for Link Power Reduction, IEEE Computer Architecture Letters, 3 (1), pp. 14–17, Jan, 2004.

Bill Moyer, Low-Power Design for Embedded Processors, Proceedings of the IEEE, 89 (11), pp. 1576–1587, Nov, 2001.

Zhigang Hu, Alper Buyuktosunoglu, Viji Srinivasan, Victor Zyuban, Hans Jacobson, and Pradip Bose, Microarchitectural Techniques for Power Gating of Execution Units, Proceedings of the International Symposium on Low Power Electronics and Design (ISLPED'04), pp. 32-37, Aug, 2004.

Kimiyoshi Usami, and Naoaki Ohkubo, A Design Approach for Fine-grained Run-Time Power Gating using Locally Extracted Sleep Signals, Proceedings of the International Conference on Computer Design (ICCD'06), Oct, 2006.

Makoto Ishikawa, Tetsuya Kamei, Yuki Kondo, Masanao Yamaoka, Yasuhisa Shimazaki, Motokazu Ozawa, Saneaki Tamaki, Mikio Furuyama, Tadashi Hoshi, Fumio Arakawa, Osamu Nishii, Kenji Hirose, Shinichi Yoshioka, and Toshihiro Hattori, A 4500 MIPS/W, 86μA Resume-Standby, 11μA Ultra-Standby Application Processor for 3G Cellular Phones, IEICE Transactions on Electronics, Vol. E88-C, num. 4, pp. 528-535, Apr, 2005.

Robert Mullins, Minimising Dynamic Power Consumption in On-Chip Networks, Proceedings of the International Symposium on System-on-Chip (SOC'06), pp. 1-4, Sep, 2006.

Michael Munch, Norbert Wehn, Bernd Wurth, Renu Mehra, and Jim Sproch, Automating RT-Level Operand Isolation to Minimize Power Consumption in Datapaths, Proceedings of the Design Automation and Test in Europe Conference (DATE'00), pp. 624-633, Mar, 2000.

Jongman Kim, Chrysostomos Nicopoulos, Dongkook Park, Vijaykrishnan Narayanan, Mazin S. Yousif, and Chita R. Das, A Gracefully Degrading and Energy-Efficient Modular Router Architecture for On-Chip Networks, Proceedings of the International Symposium on Computer Architecture (ISCA'06), pp. 14-15, Jun, 2006.

Michael Bedford Taylor, Jason Sungtae Kim, Jason E. Miller, David Wentzlaff, Fae Ghodrat, Ben Greenwald, Henry Hoffmann, Paul Johnson, Jae-Wook Lee, Walter Lee, Albert Ma, Arvind Saraf, Mark Seneski, Nathan Shnidman, Volker Strumpen, Matthew Frank, Saman P. Amarasinghe, and Anant Agarwal, The Raw Microprocessor: A Computational Fabric for Software Circuits and General Purpose Programs, IEEE Micro, Vol. 22, num. 2, pp. 25-35, Apr, 2002.

William J. Dally, and Brian Towles, Route Packets, Not Wires: On-Chip Interconnection Networks, Proceedings of the Design Automation Conference (DAC'01), pp. 684-689, Jun, 2001.

William James Dally, and Brian Towles, Principles and Practices of Interconnection Networks, Morgan Kaufmann, (2004).

William James Dally, Virtual-Channel Flow Control, IEEE Transactions on Parallel and Distributed Systems, Vol. 3, num. 2, pp. 194-205, (1992).

José Duato, A Necessary And Sufficient Condition For Deadlock-Free Adaptive Routing In Wormhole Networks, IEEE Transaction on Parallel and Distributed Systems, Vol. 6, num. 10, pp. 1055-1067, (1995).

José Duato, Sudhakar Yalamanchili, and Lionel M. Ni, Interconnection Networks: An Engineering Approach, Morgan Kaufmann, (2002).

Christopher J. Glass, and Lionel M. Ni, The Turn Model for Adaptive Routing, Proceedings of the International Symposium on Computer Architecture (ISCA'92), pp. 278-287, May, 1992.

Ge-Ming Chiu, The Odd-Even Turn Model for Adaptive Routing, IEEE Transaction on Parallel and Distributed Systems, Vol. 11, num. 7, pp. 729-738, (2000).

Luca Benini, and Giovanni De Micheli, Networks on Chips: A New SoC Paradigm, IEEE Computer, Vol. 35, num. 1, pp. 70-78, Jan, 2002.

Luca Benini, and Giovanni De Micheli, Networks on Chips: Technology And Tools, Morgan Kaufmann, (2006).

Jingcao Hu, and Radu Marculescu, DyAD - Smart Routing for Networks-on-Chip, Proceedings of the Design Automation Conference (DAC'04), pp. 260-263, Jun, 2004.

André DeHon, Unifying Mesh- and Tree-Based Programmable Interconnect, IEEE Transactions on Very Large Scale Integration Systems, Vol. 12, num. 10, pp. 1051-1065, Oct, 2004.

David Flynn, AMBA Enabling Reusable On-Chip Designs, IEEE Micro, vo. 17, num. 4, pp. 20-27, Jul, 1997.

IBM Corporation, The CoreConnect^{TM} Bus Architecture, available at `http://www.chips.ibm.com/products/coreconnect/`, (1999).

Sonics Inc., Sonics3220^{TM} SMART Interconnect IP^{TM}, available at `http://www.sonicsinc.com/`, (2002).

ARM Ltd., AMBA AXI Protocol Specification, available at `http://www.arm.com/products/solutions/AMBAHomePage.html`, (2003).

Kenichiro Anjo, Atsushi Okamura, and Masato Motomura, Wrapper-based Bus Implementation Techniques for Performance Improvement and Cost Reduction, IEEE Journal of Solid-State Circuits, Vol. 39, num. 5, pp. 804-817, May, 2004.

Charles E. Leiserson, Fat-Trees: Universal Networks for Hardware-Efficient Supercomputing, IEEE Transactions on Computers, pp. 892-901, Vol. 34, num. 10, Oct, 1985.

Charles E. Leiserson, Zahi S. Abuhamdeh, David C. Douglas, Carl R. Feynman, Mahesh N. Ganmukhi, Jeffrey V. Hill, W. Daniel Hillis, Bradley C. Kuszmaul, Margaret A. St. Pierre, David S. Wells, Monica C. Wong-Chan, Shaw-Wen Yang, and Robert Zak, The Network Architecture of the Connection Machine CM-5, Journal of Parallel and Distributed Computing, Vol. 33, num. 2, pp. 145-158, Mar, 1996.

Federico Silla, and Jose Duato, High-Performance Routing in Networks of Workstations with Irregular Topology, IEEE Transactions on parallel and distributed systems, Vol. 11, num. 7, pp. 699-719, (2000).

J. E. Smith and G. Sohi, *The Microarchitecture of Superscalar Processors*. Proceedings of IEEE, Vol. 83, No. 12, pp. 1609-1624, Dec. 1995.

. Gowan, L. Biro, D. Jackson, *Power considerations in the Design of the Alpha 21264 Microprocessor*. In CAD1998, The 35th Design Automation Conference, pp. 726-731, June 1998.

V. Tiwari *et al. Reducing Power in High-performance Microprocessors*. In CAD 1998, 35th Design Automation Conference, San Francisco, pp. 732-737, June 1998.

M. Sowa, A. Ben-Abdallah and T. Yoshinaga, *Parallel Queue Processor Architecture Based on Produced Order Computation Model*. In Int. Journal of Supercomputing, HPC, Vol.32, No.3, pp.217-229, June 2005.

A. Ben-Abdallah, T. Yoshinaga, M. Sowa, *High-Level Modeling and FPGA Prototyping of Produced Order Parallel Queue Processor Core*. In Int. Journal of supercomputing, Volume 38, Number 1, pp. 3-15, October 2006.

B. R. Preiss, V. C. Hamacher, *Data Flow on Queue Machine*. In ISCA 1985, 12th International Symposium on Computer Architecture, Boston, pp. 342-351, August 1985.

M. Fernandes, J. Llosa, N. Topham, *Using Queues for Register File Organization in VLIW*. Technical Report ECS-CSG-29-97, University of Edinburgh, Department of Computer Science, 1997.

L. S. Heath, S. V. Pemmaraju, A. N. Trenk, *Stack and Queue Layouts of Directed Acyclic Graphs: Part I*. In SIAM Journal of Computing, Vol 23, No. 4, pp.1510-1539, 1996.

H. Schmit, B. Levine, B. Ylvisaker, *Queue Machines: Hardware Compilation in Hardware*. In FCCM'02, 10th Annual IEEE Symposium on Field-Programmable Custom Computing Machines, pp. 152-161, 2002.

A. Ben-Abdallah, M. Arsenji, S. Shigeta, T. Yoshinaga, M. Sowa, *Queue Processor for Novel Queue Computing Paradigm Based on Produced Order Scheme*. In Proceedings of HPC, IEEE CS, pp. 169-177, July 2004.

J. P. Koopman, *Stack Computer*. Ellis Horwood Limited.

Harlan McGhan and Mike O'Connor, *PicoJava: A direct execution engine for Java bytecode*. in

Computer Trans., Vol.31, No.10, pp. 22-30, October 1998.

Advancel Logic Corporation, *Tiny2J Microprocessor Core for Javacard Applications*. http://www.advancel.com

A. Ben-Abdallah, M. Sarem., and M. Sowa, *DRA:Dynamic Register Allocator Mechanism For FARM Microprocessor*, in APPT99, 1999.

G. De Micheli, R. Ernst and W. Wolf, *Readings in Hardware/Software co-design*. Morka Kaufmann Publishers, ISBN 1-55860-702-1.

M. Sheliga and E. H. Sha, *Hardware/Software Co-design With the HMS Framework*. In the Journal of VLSI Signal Processing Systems, Vol. 13, No.1, pp. 37-56, 1996.

K. Kim, H. Y. Kim and T. G. Kim, *Top-down Retargetable Framework with Token-level Design for Accelerating Simulation Time of Processor Architecture*. In IEICE Trans. Fundamentals of Electronics, Communications and Computer Sciences, Vol. E86-A, No. 12, pp.3089-3098, Dec. 2003.

D. Lewis et al, *The Stratix Logic and Routing Architecture*. In FPGA-02, International Conference on FPGA, pp 12-20, 2002.

Altera Design Software. http://www.altera.com/

http://www.arm.com/products/CPUs/ARM926EJ-S.html

JazelleTM-ARM Architecture Extensions for Java Applications, White Paper, http:// www.arm.com

Oyvind Strgm, Einar J. Aas., *An Implementation of an Embedded Microprocessor Core with support for Executing Byte Compiled Java Code*, in: Proceedings of the Euromicro Symposium on Digital Systems Design, 2001, pp. 396-399.

M. Sowa, *PQP: Parallel Queue Processor with High Parallelism, Simple Hardware and Low Power Consumption*, technical Report, SLL030331, the Univ. of Electro-Communications, Sowa laboratory, June 2003.

http://ultratechnology.com/

M. Akanda, A. Ben-Abdallah, S. Kawata, and M. Sowa, *An Efficient Dynamic Switching Mechanism (DSM) for Hybrid Processor Architecture*. In Proceedings of Springer's Lecture Note in Computer Science, LNCS-3824, pp. 77-86, December 6-9, 2005.

M. Akanda, A. Ben-Abdallah, and M. Sowa, *On the Design of a Dual-Execution Mode Processor: Architecture and Preliminary Evaluation*, In Proceedings Springer's Lecture Note in Computer Science, LNCS-4331, pp. 37-46, December 1-4, 2006.

H. Suzuki, O. Shusuke, A. Maeda and M. Sowa, *Implementation and evaluation of a Superscalar Processor Based on Queue Machine Computation Model*, IPSJ SIG, Vol.99(21), pp. 91-96, 1999.

M. Sowa, *Fundamental of Queue machine*, technical Reports SLL97302 the Univ. of Electro-Communications, Sowa Laboratory, 2000.

M. Sowa, *Queue Processor Instruction Set Design*, technical Report SLL 00303, the Univ. of Electro-Communications, Sowa laboratory, 2000.

P6 Power Data Slides provided by Intel Corp. to Universities.

B. Bisshop, T. Killiher, and M. Irwin, *The Design of Register Renaming Unit*. In Proceedings of Great Lakes Symposium on VLSI, 1999.

Cadence Design Systems. http://www.cadence.com/

F. Arahata, O. Nishii, K. Uchiyama, N. Nakagawa, *Functional verification of the superscalar SH-4 microprocessor*, In Compcon97, the Proceedings of the International conference Compcon97, pp. 115–120, Feb 1997.

SuperH RISC engine SH-1/Sh-2/Sh-DSP Programming Manual. http://www.renesas.com

H. Maejima, M. Kinaga and K. Uchiyama, *Design and architecture for Low Power/High Speed RISC Microprocesor:SuperH*, In IEICE Transaction on Electronics, Vol. E80, No. 12, pp. 1539–1549, Dec. 1997.

H. Takahashi, S. Abiko and S. Mizushima, *A 100 MIPS High Speed and Low Power Digital Signal*

Processor, In IEICE Transaction on Electronics, Vol. E80-C, No. 12, pp. 1546–1552, 1997.

R. Lysecky and F. Vahid, *A Study of the Speedups and Competitiveness of FPGA Soft Processor Cores using Dynamic Hardware/Software Partitioning*. In Proceedings of Design Automation and Test in Europe (DATE'05), Munich, Germany, Vol. 1, pp. 18–23, March 2005.

A. Ben-Abdallah, *Dynamic Instructions Issue Algorithm and a Queue Execution Model Toward the Design of Hybrid Processor Architecture*. Ph.D. thesis, Graduate School of Information Systems, the Univ. of Electro-Communications, March 2002.

Advanced RISC Machines Ltd, *ARM7DMI Data Sheet*. 1994.

John L. Hennessy, David A. Patterson, *Computer Architecture A Quantitative Approach*, Morgan Kaufmann Publishers, Inc. San Francisco, California (1996).

Advanced RISC Machines Ltd, *ARM Architecture Reference Manual*, 02 September, 2001.

Gaisler Research Laboratory, *LEON2 XST User's Manual*, 1.0.22 edition, May 2004.

A. Canedo, A. Ben-Abdallah, M. Sowa, *Code Generation Algorithms for Consumed and Produced Order Queue Machines*, Master Thesis, Graduate School of Information Systems, University of Electro-Communications, September 2006.

J. Merrill, *GENERIC and GIMPLE: A New Tree Representation for Entire Functions*, In Proceedings of GCC Developers Summit, pp. 171-180, 2003.

N. VijayKrishnan, *Issues in the Design of JAVA Processor Architecture*, Ph. D thesis, University of South Florida, Tampa, FL-33620, December 1998.

D. Mattson, M. Christensson, *Evaluation of synthesizable CPU cores*, Master's Thesis, Department of Computer Engineering, Chalmers University of Technology, 2004.

2006 EDN DSP directory. http://www.edn.com/dspdirectory

CPU86 8088/8086 FPGA IP Core. http://www.ht-lab.com/

Sowa Laboratory: *www.sowa.is.uec.ac.jp*

International technology roadmap for semiconductors, Technical report, http://public.itrs.net.

Sylvester, D., Keutzer, K., A global wiring paradigm for deep submicron design, IEEE Trans. on CAD of Integrated Circuits and Systems, Vol. 19, No. 2, Feb. 2000.

A. Hemani *et al.*, Network on a chip: an architecture for billion tran- sistor era, Proc. of the IEEE NorChip Conf., Nov.2000.

W. J. Dally, B. Towles, Route packets, not wires: on-chip interconnec- tion networks, Proc. DAC, pp. 684–689, June 2001.

S. Kumar *et al.*, A network on chip architecture and design methodology, Proc. Symposium on VLSI, pp. 117–124, April 2002.

A . Pmto, L. P. Carloni, and A. L. Sangiovannin-Vincentelli, Efficient Synthesis of Netork on Chips, In ICCD 2003.

G. Ascia, V. Catania, and M. Palesi, Multi Aware Mapping for Mesh based NoC Architectures, in Proceedings of ISSS-CODES, 2004.

J. Hu and R. Marculescu. Energy-aware mapping for tile-based NoC architectures under performance constraints. ASP-DAC, pp.233-239, Jan. 2003.

J. Hu, R. Marculescu. Exploiting the Routing Flexibility for Energy/Performance Aware Mapping of Regular NoC Architectures. Proc. Design Automation and Test in Europe (DATE), March, 2003.

S. Pasritch *et al.*, Floorplan-aware automated synthesis fo bus based communication architectures, Proc.of the 42nd annual conference on Design automation, San Diego, California, pp.565 - 570, 2005.

J. Hu *et al.*, System level Point to point Copmmunication Sysnthesis Using Floorplanning informa- tion, ASPDAC/VLSI, 2002.

Ho, R. Mai, K.W. Horowitz, M.A., The future of wires, Proceedings of the IEEE, Volume: 89, Issue: 4, pp.490-504, Apr. 2001.

S. Murali *et al.*, Application Specific Network-on-Chip Design with Guaranteed Quality Approximation Algorithms, ICCAD'06, San Jose, CA, Nov. 2006.

Dash opt. tool: http://www.dashoptimization.com/.

A. Jalabert, S. Murali, L. Benini, and G. De Micheli. xpipesCompiler: A tool for instantiating application specific Networks on Chip. In Proceedings of DATE, 2004.

A. Ben-Abdallah, M. Sowa, Basic Network-on-Chip Interconnection for Future Gigascale MCSoCs Applications: Communication and Computation Orthogonalization, TJASSST2006, Dec. 4-9, 2006.

E. Bolotin, I. Cidon, R. Ginosar and A. Kolodny, QNoC: QoS architecture and design process for Network on Chip, Special issue on Networks on Chip, The Journal of Systems Architecture, December 2003.

K. Diefendorff and K. Dubey, "How Multimedia Workloads Will Change Processor Design, " IEEE Computer, Vol. 30, No. 9, pp. 43–45, September 1997.

Y. Liu, S. Chakraborty, W. T. Ooi, A. Gupta and S. Mohan, "Workload characterization and cost-quality tradeoffs in MPEG-4 decoding on resource-constrained devices," in Workshop on Embedded Systems for Real-Time Multimedia, September 2005, pp. 129-134.

B. C. Mochoki, K. Lahiri, S. Cadambi and X. S. Hu, "Signature-based workload estimation for mobile 3D graphics," in ACM IEEE Design Automation Conference, 2006, pp. 592-597.

N. Tack, F. Moran, G. Lafruit and R. Lauwereins, "3D graphics rendering time modeling and control for mobile terminals," in International Conference on 3D Web Technology, Monterey, California, 2004, pp. 109-117.

G. Lafruit, L. Nachtergaele, K. Denolf and J. Bomans, "3D computational graceful degradation," in IEEE International Symposium on Circuits and Systems, Geneva, Switzerland, May 2000, pp. 547-550.

K. Asanovic, B. Bodik, J. Gebis, P. Husbands, K. Keutzer, D. Patterson, W. Plishker, J. Shalf, S. Williams and K. Yelik. "The landscape of parallel computing research: A view from Berkley, " UCB/EECS-2006-183, December 2006, Available: http://www.gigascale.org/pubs/1008/EECS-2006-183.pdf.

J. A. Fisher, "Walk-Time Techniques: Catalyst for Architectural Change," IEEE Computer, Vol. 30, No. 9, pp. 40-42, September 1997.

nVIDIA, "Tesla GPU computing technical brief," May 2007.

D. Gohringer, M. Hubner, V. Schatz, J. Becker, "Runtime adaptive multi-processor system-on-chip: RAMPSoC," in International Symposium on Parallel and Distributed Processing, April 2008, pp. 1-7.

D. Gohringer, M. Hubner, T. Perschke and J. Becker, "New dimensions for multiprocessor architectures: On demand heterogeneity, infrastructure and performance through reconfigurability - the RAMPSoC approach, " in IEEE International Field-Programmable Logic and Applications, September 2008, pp. 495-498.

R. C. Coffer and B. Harding, Rapid System Prototyping with FPGAs: Accelerating the Design Process. Burlington, MA: Elsevier, 2006.

R. Hartenstein, "Coarse grain reconfigurable architectures," in ACM Conference on Asia South Pacific Design Automation, Yokohama, Japan, 2001, pp. 564-570.

R. Hartenstein, M. Herz, T. Hoffmann and U. Nageldinger, "KressArray xplorer: A new CAD environment to optimize reconfigurable datapath architectures, " in IEEE Asia South Pacific Design Automation Conference, January 2000, Yokohama, Japan.

S. C. Goldstein, H. Schmit, S. Cadambi, M. Moe and R. R. Taylor, "PipeRench: A Reconfigurable Architecture and Compiler," IEEE Computer, Vol. 33, No. 4, pp. 70–77, April 2000.

T. J. Callahan, J. R. Hauser and J. Wawrzynek, "The Garp Architecture and C Compiler, " IEEE Computer, Vol. 33, No. 4, pp. 62–69, April 2000.

A. Marshall, T. Stansfield, I. Kostarnov, J. Vuillemin and B. Hutchings, "A reconfigurable arithmetic array for multimedia applications," in ACM International Conference on FPGAs, 1999, pp.

135–143.

A. Alsolaim, J. Becker, M. Glesner and J. Starzyk, "Architecture and application of a dynamically re-configurable hardware array for future mobile communication systems," in IEEE International Conference on Field-Programmable Custom Computing Machines, pp. 205–214, 2000.

Xilinx, "Virtex-5 Family Overview," February 2009.

Altera, "Stratix IV Device Handbook - Volume 1," April 2009.

M. F. Jacome and G. De Veciana, "Design Challenges for New Application-Specific Processors", IEEE Design & Test of Computers, Vol. 17, No. 2, pp. 40–50, April 2000.

T. Bjerregaard and S. Mahadevan, "A Survey of Research and Practices of Network-on-Chip," ACM Computing Surveys, Vol. 38, No. 1, pp. 1–51, March 2006.

Intel, "Intel core duo processors," 2006.

Wikipedia, "AMD Multicore Opteron," March 2009.

J. Barreh, J. Brooks, R. Golla, G. Grohoski, R. Hetherington, P. Jordan, M. Luttrell, C. Olson and M. Shah, "Niagara-2: A Highly Threaded Server-on-a-Chip," August 2006, Available at http://www.opensparc.net/pubs/preszo/06/HotChips06_09_ppt_master.pdf.

F. Barat, R. Lauwereins and G. Deconinck, "Reconfigurable Instruction Set Processors from a Hard-ware/Software Perspective," IEEE Transactions on Software Engineering, Vol. 28, No. 9, pp. 847–862, September 2002.

J. Becker, M. Hubner and K. Paulsson, "Physical 2D morphware and power reduction methods for ev-eryone," in Dynamically Reconfigurable Architectures, P. Athanas, J. Becker and G. Brebner, Eds. 2006.

S. Shukla, N. W. Bergmann and J. Becker, "QUKU: A coarse grained paradigm for FPGAs," in Dynamically Reconfigurable Architectures, Vol. LNCS 6141, P. M. Athanas, J. Becker and G. Brebner, Eds. January 2006.

S. Shukla, N. W. Bergmann and J. Becker, "QUKU: A fast run time reconfigurable platform for image edge detection," in Reconfigurable Computing: Architectures and Applications, Vol. LNCS 3985, Berlin/Heidelberg: Springer, pp. 93–98, August 2006.

S. Shukla, N. W. Bergmann and J. Becker, "QUKU: A two-level reconfigurable architecture," in IEEE Computer Society Annual Symposium on VLSI, Karlsruhe, Germany, 6 pages, March 2006.

S. Shukla, N. W. Bergmann and J. Becker, "QUKU: A FPGA based flexible coarse grain architecture design paradigm using process networks," in IEEE International Symposium on Parallel and Distributed Processing, pp. 1–7, March 2007.

P. M. Heysters, "Coarse Grained Reconfigurable Processors," Ph. Dissertation, University of Twente, Netherlands, 2004.

A. Montone, V. Rana, M. D. Santambroglio and D. Sciuto, "HARPE: A Harvard-based processing element tailored for partial dynamic reconfigurable architectures," in IEEE International Sym-posium on Parallel and Distributed Processing, Miami, Florida, pp. 1–8, April 2008.

D. Kulkarani, W. A. Najjar, R. Rinker. and F. J. Kurdahi, "Fast area estimation to support com-piler optimization in FPGA-based reconfigurable systems," in IEEE Symposium on Field-Programmable Custom Computing Machines, Napa, California, pp. 239–247, April 2002.

E. Altman, B. R. Childres, R. Cohen, J. Davidson, K. De Bosschere, B. De Sutter, M. A. Ertl, M. Franz, Y. Gu, M. Hauswirth, T. Heinz, R. Bosch, W.-C. Hsu, J. Knoop, A. Krall, N. Kumar, J. Maebe, R. Muth, R. Xavier, E. Rohu, R. Rosner, M. L. Soffa, J. Troeger and C. Vick, "Final report – emerging uses and paradigms for dynamic binary translation," in Emerging Uses and Paradigms for Dynamic Binary Translation , Vol. 8441, Dagtstuhl Seminar, February 2009.

E. A. Altman, K. Ebcioglu, M. Gschwind and S. Sathaye, "Advances and Future Challenges in Binary Translation and Optimization," in Proceedings of the IEEE, Vol. 89, No. 11, pp. 1710–1722, November 2001.

G. Stitt and F. Vahid, "Hardware/Software partitioning of software binaries," in IEEE/ACM Interna-tional Conference on Computer-Aided Design, San Jose, California, pp. 164–170, 2002.

R. Lysecky, G. Stitt and F. Vahid, "Warp Processors," in ACM Transactions on Design Automation of Electronic Systems, Vol. 11, No. 3, pp. 659–681, July 2006.

R. Lysecky and F. Vahid, "A configurable logic architecture for dynamic Hardware/Software partitioning," in ACM Design Automation and Test in Europe, pp. 1048–1053, 2004.

A. C. S. Beck, M. B. Rutzig, G. Gaydadjiev and L. Carro, "Transparent reconfigurable acceleration for heterogeneous embedded applications," in IEEE Design Automation and Test in Europe, pp. 1208–1213, March 2008.

A. C. S. Beck and L. Carro, "Application of binary translation to java reconfigurable architectures," in IEEE International Symposium on Parallel and Distributed Processing, Denver, Colorado, pp. 1–8, April 2005.

A. Gonzalez, J. Tubella and C. Molina, "Trace-level reuse," in IEEE International Conference on Parallel Processing, Aizu-Wakamatsu City, Japan, pp. 30–37, September 1999.

K. C. Yeager, "The MIPS R10000 Superscalar Microprocessor," IEEE Micro, Vol. 16, No. 2, pp. 28–40, April 1996.

P. M. Heysters, "Coarse-grained reconfigurable computing power aware applications," in International Conference on Reconfigurable Systems and Algorithms, Las Vegas, Nevada, pp. 274–28, June 20061.

E. Gamma, R. Helm, R. Johnson and J. Vissides, Design Patterns: Elements of Reusable Object-Oriented Software, Addison-Wesley, 1995.

M. Gschwing, E. Altman, E. Sathaye, P. Ledak and D. Appenzeller, "Dynamic and Transparent Binary Translation," IEEE Computer, Vol. 33, No. 3, pp. 54–59, March 2000.

J. H. Pan, T. Mitra and W.-F. Wong, "Configuration bitstream compression for dynamically reconfigurable FPGAs," in IEEE/ACM International Conference on Computer-Aided Design, San Jose, California, November 2004, pp. 766-773.

Z. Li and S. Hauck, "Configuration compression for Virtex FPGAs," in IEEE Annual Symposium on Field-Programmable Custom Computing Machines, Napa, California, pp. 147–159, 2001.

A. Dandalis and V. Prasanna, "Configuration Compression for FPGA-Based Embedded Systems," IEEE Transactions on VLSI, Vol. 13, No. 12, pp. 1394–1398, December 2005.

S. R. Park and W. Burleson, "Configuration cloning: Exploiting regularity in dynamic DSP architectures," in ACM International Symposium on FPGAs, Monterey, California, pp. 81–89, 1999.

R. Hect, S. Kubisch, A. Herrholtz and D. Timmermann, "Dynamic reconfiguration with hard-wired networks-on-chip on future FPGAs," in IEEE International Conference on Field-Programmable Logic and Applications, Tampere, Finland, pp. 527–530, August 2005.

K. Goossens, M. Bennebroek, Y. H. Jae and M. A. Wahlah, "Hardwired networks on chip in FPGAs to unify functional and configuration interconnects," in ACM/IEEE International Symposium Networks-on-Chip, pp. 45–54, April 2008.

G. De Micheli, R. Ernst and W. Wolf, Readings in Hardware/Software co-design, Morka Kaufmann Publishers, ISBN 1-55860-702-1, 2001.

M. Sowa, A. Ben-Abdallah and T. Yoshinaga, Parallel Queue Processor Architecture Based on Produced Order Computation Model, Journal of Supercomputing, Vol. 32, No. 3, pp. 217–229, 2005.

A. Ben-Abdallah, Dynamic Instructions Issue Algorithm and a Queue Execution Model Toward the Design of Hybrid Processor Architecture, Ph.D. thesis, Graduate School of Information Systems, the Univ. of Electro-Communications, March 2002.

M. Sheliga and E. H. Sha, Hardware/Software Co-design With the HMS Framework, Journal of VLSI Signal Processing Systems, Vol. 13, No. 1, pp. 37–56, 1996.

D. Lewis et al., The Stratix Logic and Routing Architecture, in: FPGA-02, International Conference on FPGA, pp. 12–20, 2002.

Cadence Design Systems:http://www.cadence.com/.

Altera Design Software: http://www.altera.com/.

A. Sharma and R. Jain, *Estimating Architectural Resources and Performance for High-Level Synthesis Applications*, in: DAC 1993, The 30th International Conference on Design automation, pp. 355–360, 1993.

K. Kim, H. Y. Kim and T. G. Kim, *Top-down Retargetable Framework with Token-level Design for Accelerating Simulation Time of Processor Architecture*, IEICE Trans. Fundamentals of Electronics, Communications and Computer Sciences, Vol. E86-A, No. 12, pp. 3089–3098, Dec. 2003.

H. Maejima, M. Kinaga and K. Uchiyama, *Design and architecture for Low Power/High Speed RISC Microprocesor:SuperH*, IEICE Transaction on Electronics, Vol. E80, No. 12, pp. 1539–1549, dec. 1997.

H. Takahashi, S. Abiko and S. Mizushima, *A 100 MIPS High Speed and Low Power Digital Signal Processor*, IEICE Transaction on Electronics, Vol. E80-C, No. 12, pp. 1546–1552, 1997.

IEEE Standard for Binary Floating-point Arithmetic, ANSI/IEEE Standard 754, 1985.

F. Arahata, O. Nishii, K. Uchiyama, N. Nakagawa, Functional verification of the superscalar SH-4 microprocessor, in Compcon97, the Proceedings of the International conference Compcon97, Feb. 1997, pp. 115–120.

SuperH RISC engine SH-1/Sh-2/Sh-DSP Programming Manual: http://www.renesas.com.

C. Lee, M. Potkonjak, and W.H. Mangione-Smith, MediaBench: a tool for evaluating and synthesizing multimedia and communications systems, In 30th Annual International Symposium on Microarchitecture (Micro '97), page 330, 1997.

R. Matthew, J. Rey Ringenberg, D. Ernst, T. M. Austin, T. Mudge, and R. B. Brown, MiBench: A free, commercially representative embedded benchmark suite, In IEEE 4th Annual Workshop on Workload Characterization, pp. 3–14, 2001.

G. Kane and J. Heinrich, MIPS RISC Architecture, Prentice Hall, 1992.

A. Canedo, A. Ben-Abdallah, and Masahiro Sowa, A New Code Generation Algorithm for 2-offset Producer Order Queue Computation Model, To appear in the Journal of Computer Languages, Systems and Structures, 2007.

V.A. Patankar, A. Jain, and R.E. Bryant, Formal verification of an ARM processor, In Twelfth International Conference On VLSI Design, pp. 282–287, 1999.

K. Kissel, MIPS16: High-density MIPS for the embedded market, Technical report, Silicon Graphics MIPS Group, 1997.

L. Goudge and S. Segars, Thumb: Reducing the Cost of 32-bit RISC Performance in Portable and Consumer Applications. In Proceedings of COMPCON96, pp. 176–181, 1996.

D. Alpert and D. Avnon, Architecture of the Pentium microprocessor, Micro, IEEE, 13(3), pp. 11–21, June 1993.

Xilinx MicroBlaze. http://www.xilinx.com/xlnx/.

A. Ben-Abdallah, S. Kawata, and M. Sowa, Design and Architecture for an Embedded 32-bit QueueCore, *Journal of Embedded Computing*, 2 (2): 191–205, 2006.

A. Ben-Abdallah, T. Yoshinaga, and M. Sowa, High-Level Modeling and FPGA Prototyping of Produced Order Parallel Queue Processor Core, *Journal of Supercomputing*, 38 (1): 3–15, October 2006.

A. V. Aho, R. Sethi, and J. D. Ullman, *Compilers Principles, Techniques, and Tools*, Addison Wesley, 1986.

R. Allen and K. Kennedy, *Optimizing Compilers for Modern Architectures*, Morgan Kaufman, 2002.

A. Canedo, Code Generation Algorithms for Consumed and Produced Order Queue Machines, Master's thesis, University of Electro-Communications, Tokyo, Japan, September 2006.

A. Canedo, A. Ben-Abdallah, and M. Sowa, A GCC-based Compiler for the Queue Register Processor, In *Proceedings of International Workshop on Modern Science and Technology*, pages 250–255, May 2006.

M. Fernandes, Using Queues for Register File Organization in VLIW Architectures, Technical

Report ECS-CSG-29-97, University of Edinburgh, 1997.

L. Goudge and S. Segars, Thumb: Reducing the Cost of 32-bit RISC Performance in Portable and Consumer Applications, In *Proceedings of COMPCON '96*, pages 176–181, 1996.

L. S. Heath and S. V. Pemmaraju, Stack and Queue Layouts of Directed Acyclic Graphs: Part I, *SIAM Journal on Computing*, 28 (4): 1510–1539, 1999.

J. Hennessy and D. Patterson, *Computer Architecture: A Quantitative Approach*, Morgan Kaufman, 1990.

X. Huang, S. Carr, and P. Sweany, Loop Transformations for Architectures with Partitioned Register Banks, In *Proceedings of the ACM SIGPLAN workshop on Languages, compilers and tools for embedded systems*, pages 48–55, 2001.

Sparc-International, *The SPARC Architecture Manual, Version 8*, Prentice Hall, 1992.

S. Jang, S. Carr, P. Sweany, and D. Kuras, A Code Generation Framework for VLIW Architectures with Partitioned Register Banks, In *Proceedings of the 3rd International Conference on Massively Parallel Computing Systems*, 1998.

J. Janssen and H. Corporaal, Partitioned Register File for TTAs, In *Proceedings of the 28th annual international symposium on Microarchitecture*, pages 303–312, 1995.

G. Kane and J. Heinrich, *MIPS RISC Architecture*, Prentice Hall, 1992.

R. Kessler, The Alpha 21264 microprocessor, *IEEE Micro*, 19 (2): 24–36, April 1999.

K. Kissel, MIPS16: High-density MIPS for the embedded market, Technical report, Silicon Graphics MIPS Group, 1997.

G. Kucuk, O. Ergin, D. Ponomarev, and K. Ghose, Energy efficient register renaming, *Lecture Notes in Computer Science*, 2799/2003: 219–228, September 2003.

M. Lam, Software pipelining: an effective scheduling technique for VLIW machines, In *Proceedings of the ACM SIGPLAN 1988 conference on Programming Language design and Implementation*, pages 318–328, 1988.

J. Losa, E. Ayguade, and M. Valero, Quantitative Evaluation of Register Pressure on Software Pipelined Loops, *International Journal of Parallel Programming*, 26 (2): 121–142, April 1998.

S. A. Mahlke, W. Y. Chen, P. P. Chang, and W. mei W. Hwu, Scalar Program Performance on Muliple-Instruction-Issue Processors with a Limited Number of Registers, In *Proceedings of the 25th Annual Hawaii Int'l Conference on System Sciences*, pages 34–44, 1992.

S. S. Muchnick, *Advanced Compiler Design and Implementation*, Morgan Kaufman, 1997.

D. Novillo, Design and Implementation of Tree SSA, In *Proceedings of GCC Developers Summit*, pages 119–130, 2004.

S. Okamoto, Design of a Superscalar Processor Based on Queue Machine Computation Model, In *IEEE Pacific Rim Conference on Communications, Computers and Signal Processing*, pages 151–154, 1999.

S. Pinter, Register Allocation with Instruction Scheduling, In *Proceedings of the ACM SIGPLAN 1993 conference on Programming language design and implementation*, pages 248–257, 1993.

M. Postiff, D. Greene, and T. Mudge, The Need for Large Register File in Integer Codes, Technical Report CSE-TR-434-00, University of Michigan, 2000.

B. Preiss and C. Hamacher, Data Flow on Queue Machines, In *12th Int. IEEE Symposium on computer Architecture*, pages 342–351, 1985.

R. Rau, Iterative modulo scheduling: an algorithm for software pipelining loops, In *Proceedings of the 27th annual international symposium on Microarchitecture*, pages 63–74, 1994.

R. Ravindran, R. Senger, E. Marsman, G. Dasika, M. Guthaus, S. Mahlke, and R. Brown, Partitioning Variables across Register Windows to Reduce Spill Code in a Low-Power Processor, *IEEE Transactions on Computers*, 54 (8): 998–1012, August 2005.

H. Schmit, B. Levine, and B. Ylvisaker, Queue Machines: Hardware Computation in Hardware, In *10th Annual IEEE Symposium on Field-Programmable Custom Computing Machines*, page

152, 2002.

M. Sowa, A. Ben Abdallah, and T. Yoshinaga, Parallel Queue Processor Architecture Based on Produced Order Computation Model, *Journal of Supercomputing*, 32 (3): 217–229, June 2005.

G. Tyson, M. Smelyanskiy, and E. Davidson, Evaluating the Use of Register Queues in Software Pipelined Loops, *IEEE Transactions on Computers*, 50 (8): 769–783, August 2001.

D. Wall, Limits of instruction-level parallelism, *ACM SIGARCH Computer Architecture News*, 19 (2): 176–188, April 1991.

M. Wolfe, *High Performance Compilers for Parallel Computing*, Addison-Wesley, 1996.

J. Zalamea, J. Llosa, E. Ayguade, and M. Valero, Software and Hardware Techniques to Optimize Register File Utilization in VLIW Architectures, *International Journal of Parallel Programming*, 32 (6): 447–474, December 2004.

Index

A

abstract syntax tree (AST), 66,
address pointer hardware (APH), 112
address window (AW), 8, 114
area (ARA), 13, 124, 125
area optimization (AOP), 103–105
ASIC, 2, 14, 122–123
automation algorithm, 1, 2
AVAILABLE signal, 102, 120

B

balanced (BLD), 13, 14, 124
barrier, 8, 13, 113, 123, 126
battery life, 17, 28, 29, 35, 37
binary translation (BT), 152, 153, 155–156
Booth, 11, 118

C

circular queue register (QREG), 7, 8, 13, 15, 108–111, 114–115, 117–119, 122–126
CMOS, 19–21, 23, 37, 54, 129
coarse grain parallelism, 61, 128
communication architecture, 1–2, 4–5
communication protocols, 1, 5, 6, 30, 31, 157
configurable logic blocks (CLB), 146–148

continuous data types, 128
control unit (CTRL), 97, 99–100
CoreConnect protocols, 4
covop, 64, 73–75, 84, 93, 111–112
CPI, 14, 107, 122
custom multiprocessor, 3

D

data memory, 7, 83, 102, 103, 108, 111, 123, 131, 138
decode buffer (DB), 84
decode circuits (DC), 84
decode unit (DU), 7, 83, 88, 93–95, 97, 104, 113
Default-backup path (DBP), 42, 47
denormalization, 11, 118
design flow graph (DFG), 5, 6, 63, 72
directed acyclic graphs (DAG), 62–69
dominance of iterative kernels, 128
DSP, 2, 18, 22, 32, 106
dual-execution processor (DEP), 79–82, 84-86, 88–89, 91–92, 94, 97, 100–106
dynamic instruction merging (DIM), 148–152

E

EDA tools, 12
embedded systems, 2, 61, 107, 129, 143,